DATE DUE

MOST SPLENDID
OF MEN

MOST SPLENDID OF MEN

Life in a
Mining Community 1917-25

HAROLD BROWN

BLANDFORD PRESS
Poole   Dorset

First published in the U.K. 1981

Copyright © 1981 Blandford Press Ltd.,
Link House, West Street,
Poole, Dorset, BH15 1LL

British Library Cataloguing in Publication Data
Brown, Harold
  Most splendid of men.
  1. Coal mines and mining – England – Silverdale,
  Staffordshire
  I. Title
  622'.33'0924   TN808.G62S/

          ISBN 0 7137 1107 8

Typeset by Tonbridge Printers Ltd.
Peach Hall Works,
Tonbridge, Kent.
Printed and bound in Great Britain by
Biddles of Guildford

Contents

Acknowledgements

It is impossible to mention by name the many people who have contributed information during my research. I acknowledge with gratitude their courtesy and assistance. To the following people who willingly afforded me their time and the benefit of their experience, I am especially grateful.

The Rt. Hon. Lord Robens of Woldingham, PC, DCL, LLD
Mr John Golding, MP

NATIONAL COAL BOARD, STOKE-ON-TRENT

Mr J. R. Hunter, Area Director, Western Area: Mr K. Eardley, Head of Secretariat: Mr J. H. Williams, Area Staff Manager: Mr W. Wilcox, MBE, Area Production Manager: Mr J. Hebblethwaite, MBE, Deputy Area Production Manager: Mr H. Newton, Superintendant, Mines Rescue Department; Mr J. A. Belcher, Colliery Manager, Silverdale; Mr J. Beardmore, BEM, Mechanical Engineer, Silverdale Colliery; Mr Robert Wood, Safety Officer, Silverdale Colliery; Mr Tom Sanders, Special Duties Officer, North Staffordshire Area, based at Kimball Training Centre; Mr G. Jebb, Training Officer, Silverdale Colliery; Mr C. Sutton, Official at Silverdale Colliery; The late Mr L. Bebbington, Silverdale Colliery; Mr A. Boot, Silverdale Colliery.

My thanks go equally to the Public Relations Departments, National Coal Board at Cannock and at Lowton; The late Councillor J. Evans, MBE, Burslem, ex–Lord Mayor, Stoke–on–Trent; Councillor A. Moran, Longport, ex–Lord Mayor, Stoke–on–Trent; The late Mr C. Davenport, Lord Mayor's Secretary, Stoke–on–Trent; Mr K. D. Miller, DPA, FLA, City Librarian, Stoke–on–Trent; The late Mr H. Minshall, May Bank, Stoke–on–Trent; Mr C. Bradley, Head Librarian, Newcastle–under–Lyme; Mr B. Bemrose, Curator, The Arts Centre and Museum, Newcastle–under–Lyme; Mr N. Beckett, News Editor, *The Evening Sentinel,* Hanley, Stoke–on–Trent.

H.B.

Foreword

by

The Rt. Hon. Lord Robens of Woldingham, PC, DCL, LLD

Two centuries of coal mining in the United Kingdom have produced a
rich reservoir of human experience. It has embraced all facets of
human life, tragedy and joy, hardship and courage, a comradeship
that binds men and women in their mining village close together, the
halcyon days of pre–1914 and the misery of the long years of the
1920s and the 1930s; the organisation of the men into their unions,
the struggles, the strifes, the strikes, the imprisonments, the riots,
the lock outs—all these, and much more besides, are in this vast
reservoir.

So rich is the experience of the individual miner that almost every
single pitman could write a story, and it would be *his* story, it would
be full of interest.

This book is essentially the life story of a man who, as a boy, felt the
economic pressure which decided his life for him. Had the wages
where he worked as a boy in the Co–operative store been higher than
the wages he received at the local pit, he would probably have
continued to work for the Co–operative Movement, and who knows
what position he would have occupied in that great body. As it was,
the pit offered more money and family income was very important;
and so our author entered the pits. His encounters drawn from real life
cover the whole field of a miner's experience.

For some ten years I had the privilege of being the Chairman of the
National Coal Board and there is scarcely a pit in the United
Kingdom that I have not travelled underground. I have spent many,
many hours with the mining community and for fifteen years
represented a predominantly mining area in the House of Commons.

All this experience made me realise, perhaps more than I had anticipated, the quality and the character of the miner; solid in his friendship, frank in his conversation, a humorous turn of tongue, and a deep devotion to the welfare of the children and the old folk within the mining community. The common dangers and hazardous nature of the mining operation bound miners together in the dedication of performing their task, having in mind the safety of the whole mine. Through the generations it produced first–class workpeople and only those who have seen miners at work will ever understand the skill with which they prise the coal out of the earth, resistant to their presence.

This is a book which the wider reading public will enjoy and all those who have been associated with mining will be able, very largely, to re-live their own life's experience once again. Books such as this add to the treasure-chest of knowledge of this industry which was Britain's only indigenous source of wealth. The greatness of Britain was based on the efforts and on the labour of the mining community. Without that wealth, that source of energy, Britain could never have become a leading industrial nation in the world. Times change since energies like oil have come on to the market, and they have had a profound effect upon the mining industry. But the mining industry has not stood idly by, it has perfected new techniques, and mining today has very little relation to the mining of the days about which Mr Harold Brown writes.

That is why it is very important that the passage of time should not lose the history of the mining folk and we can only be grateful to the author for inviting us to travel down memory lane with him, sharing his experiences from youth up. It will also make those who are not familiar with mining grateful to the mining community who worked hard and long in the most difficult conditions to produce the wealth that made Great Britain great.

Introduction

by
John Golding
Member of Parliament for Newcastle–under–Lyme

This is a most splendid book about 'the most splendid of men'—
the miners of North Staffordshire. It is an account of Mr Brown's
experience as a miner between the years 1917 and 1925. From it we
learn of the hardship faced and the courage shown by men
underground, many of whom are still living among us today. 'In
every way,' he says 'it was a struggle for survival.'

One scene put the final seal of authenticity on to this book for me.
One morning when two men arrived to see me quite separately, Mr
Brown began talking to Mr Tom Sanders. Mr Sanders is typical of so
many North Staffordshire miners; he is serious, intelligent, straight-
forward and, above all, he is of great natural dignity. As they talked
about pits long since closed, about the exact day on which the gob fire
took place, and about Mr Brown's 'feyther' (much of the book is about
the relationship between father and son), so Mr Brown's face became
alight with enthusiasm and affection.

This is indeed a book written with a true affection for its characters.
It is clear that although chance took Mr Brown away, first from the pit
and then from the village of Silverdale, he has never left either in
spirit. On his first day at the pit Mr Brown's father said to him, 'You
will find life very hard in the pit, but it will make a man of you; you
will find fine characters among miners—men who work hard for their
wages and observe the pit regulations. Such men are brave and
courteous, they are second to none among working men.' It is clear
that the years in the pit have left their mark on Mr Brown.

INTRODUCTION

Mr Brown is so obviously a man of great character, drive and humanity. Had he not left Newcastle–under–Lyme, he would almost certainly have become one of its leaders, perhaps a Councillor, a Magistrate, and perhaps, the best anyone could ask, become the Member of Parliament for Newcastle–under–Lyme. Now, with this book, we have been more than repaid for this loss. Had Mr Brown remained he may have become accustomed to the slow change, so hard–fought for both in working and in living conditions, and so may not have written so vividly of the past.

As I have said, many of the men who were working with Mr Brown are still with us and it is one of the pleasures of my job to enjoy sitting before a good coal fire, listening to old miners talking of days gone by, of the Minnie Pit Disaster in 1918, when one hundred and fifty–five men and boys were killed; about the Diglake Flood Disaster when seventy–seven men drowned in that pit. As Mr Brown's father remarked . . . 'Tragedy is part of our lives.' In those times there were more bad days than good ones. Such, however, is the character of these 'most splendid of men' that, despite being crippled by injuries and by 'the dust', there are always likely to be more smiles than frowns, more tolerance than bitterness as they talk of the past.

Mining is still one of the most difficult jobs to be done. When first I went down the pit at Silverdale with my old friend, Sid Fox, my eyes were opened to the hardship faced by coalminers. Men are still, as in Mr Brown's day, relieved to come up from Silverdale Pit and smell the fresh, clean air blowing on to their faces as it comes in over the Cheshire Plain from the sea. They, like him, still face the constant contrast between darkness, ugliness and drudgery underground and the beauty of the surrounding countryside. Many of these present–day miners share Mr Brown's love of music; his passion for music is so vividly brought out in the book.

When I leave Silverdale to go to London on Monday morning at about six o'clock and I pass miners in the street, I feel grateful that I am not having to share the sinking feeling which many of them must suffer at the thought of another week at the pit. Now I am grateful to Mr Brown for having written so vividly of his experiences and for paying testimony to those miners I am proud to represent—the most splendid of men.

Preface

Those men and women who have taken the trouble to make records of their discoveries during their research and study have rendered a great service to society. A host of people have made personal sacrifice in order that such records could be made available for the benefit of future generations. With the passing of every decade more and more information and colour is added to this treasure store of history; every fresh discovery reminds us that only a part of history has been written down. There is always the history we know and the history we do not know; the discovery of the Dead Sea Scrolls was evidence enough of this. So many events and changes of the long, distant past have been obliterated by the ravages of time, and with their loss has disappeared the concomitant mode of domestic life and traits of human behaviour in the context of evolution. It is my belief that the period through which I lived as a boy and the events which I was able to observe as a growing young man were sufficiently important for them to be included in any record of Social History. The community of which I was part endured hardship and suffering with patience; they faced extreme difficulty with a cheerfulness of heart and my admiration for them prompted me to make my own record. The stark events of this period have been listed in history books, but perhaps young men and women of the future may want to look a little more closely at the detail. As they ponder and wonder what their forebears were like and how we grappled with the difficulties, they may find in my story the impact that events of history had on this close–knit mining community and how men and women shared the burden of responsibility as they faced the challenge of their time.

This book is a true story of events which come within the orbit of my own experience; no claim is made to literary style, no attempt has been made to display subtleties in language or refinement of expression. Only the expression of everyday existence would be appropriate to present the true colour of the rugged routine contained within the daily lives of those men and women as they slaved to provide food and clothing for their children, to acquire a mere survival level for themselves and, at the same time, to maintain some degree of self–respect. When I have described to an audience the conditions of life which existed for us some fifty years ago, sometimes credulity has become strained. I have made every effort to present the ordinary, pedestrian mode of speech and this has involved the use of the colourful colloquialisms which exist in Staffordshire and over the borders of the adjoining counties. This form of speech is most articulate when people are at work together, but it is also used on the domestic hearth, during sporting events and between individuals generally. It is a warm and friendly attribute of social intercourse and communication and I am sure that it has contributed to the cheerfulness and warmhearted attitude to life by which North Staffordshire people are known. Indeed, as a very young boy I knew some of the older inhabitants who were unable to express themselves verbally except by the use of this unique means. The words should present no difficulty as they appear in the context of my story.

Only the miner himself knows the feeling as he enters the cage and dangles at the end of the pit–rope over the deep, open shaft below him before he is plunged downwards, then again as he is drawn up from the world of darkness into light, fresh air and space. The miner is always aware that only the strength of the pit–rope saves him from a terrible death. During the years in which my story is set, there was one word which dominated the miner's existence once he had stepped from the cage at pit–bottom. That one word was 'DANGER'. Low roof, space restriction on each side, steel ropes under great strain swinging from floor to roof, runaway waggons and loads, a sudden fall of roof, the presence of gas, the fearful possibility of sudden explosion and a dozen other hazards; all these constituted the general climate of danger underground. Colourful phraseology cannot convey to the reader a true picture of these conditions at the beginning of the twentieth century. The chest disease caused by heavy concentration of coaldust, the eye complaint brought on by years of straining to see under the glimmer of the Davy lamp, day–to–day fatalities, widespread disability through accident, permanently bent backs and distorted bodies due to working in cramped and unnatural positions; all these revealed the price which miners paid in order that coal could be brought to the surface. Added to this there was the humiliation of gross underpayment.

This book has not been written expressly to accentuate the bad conditions underground during the period in question, but rather to present to the reader the attitude of miners towards their unsalubrious surroundings and, even, to draw attention to the high degree of responsibility which every miner displays toward his workmates. The strict regulations and the exercise of personal discipline on the question of contraband is indicative of the fact that every miner shares the responsibility for the safety of the whole pit. What claimed my attention and created my admiration was the irresistible impulse in a miner to go to any length, no matter what the risk, to save his mate was suddenly encountered. There is an impelling force which drives a miner to go to any lenght, no matter what the risk, to save his mate from becoming another victim of the common enemy overhead, the enemy of *weight*. Underground men seem to abandon all mean ideas and consideration of personal safety, replacing them with a more idealistic form of conduct. I have no doubt that this excellent behaviour exists among other groups of workers where the occupation is difficult and hazardous. It has been my aim to relate the intensity of this fine quality in human behaviour which I observed from day to day during my own working days underground. My enquiries have led me to believe that this tradition exists among miners everywhere and that its depth in the human heart and mind is commensurate with the depth of the black seam itself. The observant reader will recognise this admirable feature as he or she proceeds with the story and will see how this high standard of conduct seeps through into family life and into community relationships on the surface. I have emphasised the point that the underlying ethics of this conduct have become the foundations of communal life, especially in the daily concern for the very young and the very old.

It is not difficult, even for those with no experience among miners, to understand that a change takes place as men are lowered from the surface down through the crust of the earth to the deep workings of the pit. The physical environment changes and the men find themselves in another world. There is evidence of this change from the moment the miner leaves the cage, walks through pit–bottom and enters the utter blackness of the roads leading to distant workings and to the coalface itself; he is now in a world of inimical dimensions. Words are not sufficiently powerful to describe the atmosphere of this underground world and because of this I offer no apology for the few swear-words which appear. To spare the sensitive reader any offence I have omitted the more forceful adjectives; I trust that the atmosphere has been captured without their use. The violence of the coalface and indeed of pit work in general, often creates the necessity for violent language. Just as the nasty taste of some medicines is tolerated in anticipation of the relief they may bring, the miner working under stress uses whatever words fall to his tongue if they afford him any emotional outlet or relief for his inner frustration, though there are many miners who do not swear at all. The coalface defiantly challenges the collier for it has a thousand ages and all the might of nature in its favour; it resists all the attempts of the face–worker to extract it from between the upper and lower layers which have pressed it and preserved it through the long centuries till now it has become a precious commodity of modern life. No–one has the right to deny the face–worker any means of relief when he is dealing with such an opponent as the tightly wedged–in seam; anyone with experience in the dark world of the coalface would not grudge the miner the use of a few naughty words. The reader will observe that those men who influenced my life underground were men of high moral character. They were men of few words, they did not make useless gestures and their lives seemed to include a common feature inasmuch as they appeared to be devoid of degrading influences. Even under stress they possessed a quiet poise and their example has remained with me as my mental and moral talisman throughout life. All the aspects of pit–work served to shape my character and influence my manhood, but it was the generous attitude of the men towards each other which created within me a deep and lasting respect for them.

It is necessary for the reader to keep in mind the fact that I was a mere boy of fourteen when, like thousands of other boys, I was given a pit–lamp and ordered to go and start work in a certain part of the pit. My story betrays the fact that those early days were tinctured with the presence of fear in my mind. I had been out of school only one year and at that time our whole district was still suffering and sorrowing under the pall of horror which had descended upon us only two years before my first day down the pit. This was the tragic Minnie Pit Disaster in which one hundred and fifty–five men and boys met their death on the same day. In addition I was accustomed from early infant days to hearing constant reference in domestic conversation to the Diglake Flood Disaster when seventy–seven men were drowned in a nearby pit. So, it was very natural for a small boy to be a little afraid. Today

14

my most precious possession is the Rescue Brigade medal awarded to my father. These facts will help the reader to appreciate more deeply the significance of that first ride down the shaft, the display of courtesy and human kindness and the impact it had on my young mind. That incident laid the first stone in the foundation of my philosophy.

This book is not a stylish novel; it is a tribute to miners who displayed a standard of conduct which defies any attempt to translate it in terms of ordinary language. It is also an attempt to express my own gratitude to men and women who possessed a healthy attitude towards the normal function of 'going to work' and made each other happy by their cheerful acceptance of harsh conditions and uncertain prospects. They formed a community imbued with an indomitable spirit and a natural resilience which made them able to recover quickly from the effects of tragedy and difficult conditions. To me their character has always appeared as a becoming accompaniment to the charm and fragrance of the surrounding hills and woodlands.

I trust that the reader will search below the verbal exchanges on the domestic hearth, on the coalface and during the day–to–day activities in street, shop and office. It is my hope that he or she may recognise the underlying influences at work in society during that period in history and may, perhaps, detect among the events I have described, the cause of some of the symptoms which appear in society today.

As long as men descend to great depths into the earth there will always be danger present. To dangle at the end of a pit–rope today over the deep, dark shaft is only one degree less dangerous than it was fifty years ago, but the miner now knows that the pit itself is a much safer place in which to work. I salute all those who have contributed to this splendid transformation during more recent years from National Coal Board, right down through the hierarchy of HM Inspectors of Mines, Pit Managers and officials, down to the effort made by the single individual. I look to the day when the miner will be given full recognition for his skill and courage. To mining communities everywhere I offer this book as my tribute to their endurance and exemplary conduct.

Harold Brown
Summer, 1980

15

The Early Days

1 Leaving school

It was late afternoon on a Friday and my father had just finished his meal after returning home from work at the pit. He was still in his pit–dirt for pit–head baths had not yet been introduced, not even for officials. It was the usual practice for a miner to wash his hands, sit down and eat his meal and then take a short nap before the bigger task of getting his pit–dirt off. There were four miners in our family, but my two elder brothers had volunteered for the army and had been sent to the front line in France. Only a few weeks before we had received the sad news that the younger one of the two had been killed by a German shell. The fourth miner was our cousin Ralph; his father had been killed in the pit and my mother had taken pity on the boy, bringing him into our family to live as one of us and so avoid the lad being taken into the workhouse. Ralph worked at Leycett Pit; he was asleep upstairs for he was on nights this week. I had three sisters all older than I was and there were two brothers below me in age. So our complete family consisted of eleven members.

The normal, pedestrian tenor of conversation was suddenly halted by the sound of footsteps coming up our entry, the high narrow passageway which connects the back of the terraced houses to the front pavement in the street. We three young boys sat on small wooden stools before the fire; we all looked up at my mother who moved toward the door and with a questioning tone remarked, 'Who can this be?'

Five pairs of eyes turned to the old family clock high up on the wall as if to try and associate the time of day with the firm footsteps and the articulate rat–a–tat on the door.

My father seemed unconcerned, but said dryly, 'It's shoes, it inna clogs, so it's no–one from th' pit.'

The entry, being so high, amplified every sound and even the sound of shoes could be heard indoors, every footfall like a muffled drumbeat. My brother next below me in age looked at me enquiringly as we waited for my mother to open the door. She tucked in the corner of her working apron to make sure that the visitor would see that she had on a fairly decent skirt. To my brother's glance I replied in little more than a murmur, 'It conna be th' beggar, he dunna come till after about half past five and it inna that time yet. I wonder who it is?'

We were all familiar with the habits of our regular beggar and we knew that his footwear was so worn and his footsteps so light that the steps we had just heard could not be his. My mother came back into the living–room bringing a visitor; a blush of embarrassment came over my face as I recognised our headmaster.

'It's Mr Ellams, George. He wants to have a few words with us.' My father rose to his feet and we three boys, having been trained to 'keep out of the way' when visitors arrived, moved to leave the room.

'Hello George' said the headmaster as he shook hands with my father. 'Oh . . . can Harold stay for it concerns him?'

Fear mounted in my mind as my eyes met a sharp questioning look from my father and my thoughts raced through the events at school that day in an attempt to recall anything which I may have done to justify a complaint from our headmaster. I obeyed my father's nod which indicated that I could resume my seat on the stool. Frank Ellams and my father met very regularly on the bowling green so the seeming familiarity between them was in order.

'Well, Frank', said my father giving me another not–too–friendly look. 'What's he been up to?'

The headmaster laughed and in only a few words made it clear that he wanted my parents' permission to include my name among those suitable to enter for the scholarship examination. Turning to my mother he said, 'Mrs Brown, your Harold is a bright lad and I believe that he has a fair chance along with a few others to pass this examination to take him on to the Orme Boy's School at Newcastle. We are afforded only a handful of places in this village and I think your boy should be given the opportunity. If he does not take it, I shall enter a less able boy.'

My mother looked straight into my father's eyes, but she remained silent; she left the decision to him. Again my father looked across in my direction, but his countenance had changed and I realised that he was now pleased.

'Frank,' he replied after clearing his throat, 'Frank, you know our circumstances. We have had two lads over in France almost since the very beginning of this war and it has taken every penny we have been able to scrape together to send out food parcels to them every week.

God alone knows how many of the parcels they receive, but it's the least we can do to keep sending them out so that those lads can have a bit of real snapping [food, nourishment] sometimes. Our Gertie is a cripple and does not support herself at all. The other two girls are at work, but they earn only a pittance for their effort. You know what wages are for girls around here. Ralph just about keeps himself with his board money. There is only my money coming in to support this home, keep our Gertie and these three young ones and buy clothes for the other two girls who work away. Now, we *need* our Harold's money and we have looked forward to the day when he could earn a shilling or two. I know that he is hardly eleven yet but soon he will find a part time job and that will help his mother. Our Harold go on to Secondary School, Frank, or to High School? Oh no, we could not afford it. People like us can't afford the new clothes and other things he would need for *that* kind of school. It inna that I would not like him to have th' chance, but it just is not possible for us to meet what it would cost. No, Frank, we just *need* his money when he's old enough to start work.'

Silence followed and during this brief interval I wondered why I was being included in the list of boys considered suitable for scholarship examination. I suddenly recollected that our class teacher, Mr Buckley, had praised me for asking certain questions and 'VG' and 'EX' were appearing more frequently at the end of my work in my English exercises. But I was aware that the best scholar in our class was Jimmy Baddeley and that there were several more clever than I was. My father seemed to be gazing at some object in the far distance through the blazing coal in the fire–grate. My mother remained silent, but her eyes were troubled and she wore lines across her brow. Again Frank Ellams gave an appealing look to my mother, then turned to my father.

'Look George, do you realise that you are taking this lad's chance away in life. *Of course* it will mean sacrifice, but he's worth it, he's a worker.'

My father's eyes lit up and he threw a quick glance across the room to his bowling club associate, an expression of pride spreading across his face.

'Ah Frank, he's a worker owraight, he's a Brown inna he? All Browns are workers, thaiy knowst that Frank.'

'It's a great pity for this lad of yours deserves to go on and remain at school, but I can't make you let him take the examination.' With disappointment written across his face the headmaster re–traced his steps down our entry and *that was that*.

I had been given no say in this important matter at all and my troubled mind began at that very moment a long, long period of resentment as I, even at that tender age, realised that because we were poor and because the war had taken away my brothers and their contribution to the family purse, I was being deprived of the opportunity to LEARN, the very thing in life which gave me satisfaction. I enjoyed being at school, notwithstanding the rigour of stern discipline and the fight for survival in the schoolyard among some three hundred and fifty tough boys mostly from mining families.

The headmaster's footsteps had hardly disappeared when a very gentle knock was heard on the entry–door. We heard the latch being lifted and saw the familiar shadow standing in the small passage which connected our living–room with the ouside wash–house. We all knew that it was the beggar making his regular Friday call.

'It's th' beggar,' said my mother softly as if to make the tone of her voice match the polite gentleness of the old man's knock on the door. In our village we saw several beggars each week. Some were women, and all had their own particular days of coming round. All except our own special beggar, now standing waiting for his weekly penny, tried to earn their coppers by singing. The impression each one made on my mind has remained firmly stamped throughout life.

The beggar waited and as my mother fumbled in her purse for the penny I jumped up and said, 'Can I give it to him please?' For some reason I derived a certain pleasure from handing the old man his penny. He looked so sad and always I tried to smile as I handed him the coin. His coat was green with age, his old shoes worn right down and he shuffled along rather than walked. But, once, his eyes had met mine as we passed each other on the pavement and I suspected that he tried to say something to me in that glance; it was as though my inner compassion towards him had got through where words would have been too clumsy. This old man called only at certain houses; it seemed that economy in footwear and effort forced him to call only where there was bound to be response. I knew that my mother could hardly afford this penny, but she never failed this old man except during strikes, though at such times the beggars understood and kept away.

This old man relied upon the generosity of poor people in order to live. I tried in my simple way to understand the contradiction. My parents could not afford to let me continue at school but they could afford their contribution to help relieve an old beggar–man.

Only a few weeks after the visit of my headmaster to our house, my mother hurried towards me in the street as I returned from school at teatime. She had on her 'rough apron' made from a sugar sack and she made it quite clear that whatever it was she wanted to say to me could not wait till I was indoors. She was breathless because of her hurry.

'Harold, come on, get thy face and hands washed and tidy thy hair. Hambletons want a part–time shop–lad and thee must get down there as quick as you can and get th' job.'

I obeyed without a word, but my mind was busy wondering what this sudden change in routine would mean, even if I did get the job. It was taken for granted, at least in my mother's mind, that I was capable of doing the work for as I turned and raced for home she called out to me, 'Dunna forget now. Let Mr Hambleton know that thou hast gumption.'

Hurriedly I washed my face and parted my hair with my finger–nails and within ten minutes I was standing before the owner of this very large grocery store. He was a short man with a beard. All his family were employed in the business during peacetime, but two of the sons had gone to the war, like my own two brothers had done. There was a bakery at the back where one man struggled to bake all the bread and confectionery. Mr Hambleton knew me as an occasional customer, but most of our groceries came from the 'Co–op' where there was so much dividend in the pound. He seemed not to think twice for he engaged me right away at the wage of three shillings a week plus a penny for myself.

In common with all closely–knit communities at that time, there were a few dwellings which could be described as 'below average' and I felt sorry for those who were obliged to occupy them; but as a general rule, most houses were spotless in appearance as were the domestic habits and ideas of the occupants. Like many, many others, our house sparkled at every corner and turn. Throughout the whole area extreme cleanliness was in evidence. Outside, window–sills and pavements were washed daily and it was obvious that this religious observance and application, this constant washing of doors and windows was necessary in order to combat the dirt in the atmosphere which came from the chimneys of pits, potteries, brickworks, forges, furnaces and steelworks.

So, from a very early age I had been trained to scrub floors, polish furniture, cutlery and kitchen utensils. Everything had to glitter, tables had to be made white with scouring—sand, pavement, back and front, washed down daily; this wash was followed by the cleaning out of the fowl—pen and the pigsty and if there was any time left over, there was always the garden waiting. This had been my routine for several years. I recollect that soon after my ninth birthday my mother announced with a strong tone of command in her voice, 'Harold, thou art over nine now. It is time for thee to be shown how *to be responsible* for Sunday dinner.' Like all my brothers and sisters I had begun this training from a very early age; at the tender age of five on Saturday evenings we sat on stools preparing the buckets of vegetables for Sunday dinner. During this chore we practised the singing of our Sunday School hymns. A large family required a lot of vegetables and without my mother's expert organisation she would have floundered. There was a lot of hard and heavy work at Hambleton's shop, but I came to appreciate that my mother's training had given me a great advantage and I found some of the more menial tasks to be little more than child's play. I resented some of my mother's strictness and insistence on every speck of dust being removed; it *was* hard for a lad so young, but I did not know just what a worthwhile and useful investment my mother was making on my behalf. What she taught me stayed with me throughout life and proved to be my most useful acquirement.

However, most of the work at Hambleton's shop was very heavy indeed. The shop was situated on a corner, the side of which lay on a steep slope of Crown Street. The warehouse was on this slope and my job was to replenish the sacks of various cattle—food, fowl—food and pig—food each day as they were required. Then a constant supply of sacks of flour, sugar, soda, rice, peas, lentils etc. had to be maintained. It was a great strain on me as I struggled with these heavy sacks of goods on that steep gradient. One day as I pushed and mauled with a heavy sack of flour on the slope a passing pedestrian came to my aid and remarked with some indignation, 'Hambleton shouldna expect a bit of a lad to do this. Why thy truck itself is heavy enough without that sack of flour.' I fully agreed with the man.

The shop hours were long, and I was allowed to go home at eight

23

o'clock on Monday, Tuesday and Wednesday. Thursday was half–day closing. On Friday I had to deliver groceries on my heavy truck two miles away at Keele and to farms on the other side of the village, climbing steep, exposed hillsides. In the winter this was difficult work finding my way in the dark only by the flicker of a small lantern. Saturday night could have meant my going home as late as ten o'clock for the shop remained open as long as customers came in. I started work at 7.30 am each day except Sunday and did an hour's work before I went to school. During our lunchtime break from school I worked half an hour at the shop and I estimated that I was at work at Hambleton's for every hour that I was not in bed or at school. I earned my three shillings a week. I was proud to hand my wages to my mother for I knew that I was making a contribution to the housekeeping. Even the penny for myself was given up to my mother. Somehow I seemed not to object to this pennyworth of Sacrifice for I was deeply aware of my mother's need for help with housekeeping money and I kept saying to myself, 'I am now helping to pay my way and that extra penny will mean that I am paying the old beggar each time he calls.' I derived great pleasure from this weekend procedure of handing money to my mother and I did not mind the work either for, although it was a strain, it let me off chores at home. These were now performed by my younger brothers. One benefit was that in the winter the shop was warm, whereas at home it was not always possible for me to obtain a position near to the fire with such a large family sitting around reading, playing cards or sewing. Also, at the shop, by using stealth, there was the occasional morsel of food—tiny bits of cheese, a few grains of wheat, perhaps a fig or a date from the box when no one was looking. One feature of these early days of my life has always remained fixed in my mind; the vividness of the memory has never diminished—I was always hungry.

Two years passed quickly, for I was so hard at work the time flew by. The day I was thirteen my headmaster gave me a blue form, my certificate of attendances; he had a sad look in his eyes as I left school on that day. I could have remained till I was fourteen, but my parents had indicated that they wished me to leave. This was permitted in those days if a boy had a job to which he could go. I had a job already for Mr Hambleton had agreed to employ me at the full–time rate of six shillings a week. That did not last long, for after only a few weeks as full–time worker my mother addressed me with a serious countenance, 'There's a job at th' stores; the "Co-op" are looking for a smart shop–lad. I have already towd them that thou art suitable.' Again I was successful. No doubt the 'Co–op' knew that I was already half–trained by reason of my work at Hambleton's and they perhaps took into consideration the size of our weekly grocery order. My wages were now ten shillings a week. I was now, for the first time in my life, in receipt of pocket–money. I enjoyed the work for now my hours were more reasonable; Trade Union activities were increasing and their influence was beginning to be felt, even in the matter of shop–hours. Immediately I turned to night school, studying Double–entry Book–keeping. I needed no persuasion to get back into a school desk and I applied myself seriously to my studies. I added Commercial English to my list and began to set my sights on the possibility of becoming a grocery manager when I reached manhood.

My mother's strict domestic training had equipped me well and, with two years of experience at Hambleton's, my prospects were quite fair. Mr Hambleton had allowed me to continue wearing my clogs at work for I had to have them for school. The only time I had ever worn shoes was for Sunday School and Chapel attendances. There was nothing unusual about a person working in clogs in other occupations than mining. Some girls wore them in the fustian mills and many women, including my own mother, performed their household duties in clogs; clogs were warmer and very much more economical than shoes. So, when I was informed at the 'Co–op' on my first day there that I should have to wear shoes or boots and *not* clogs, I felt that I had taken a step up in the world. All went well and I experienced a glow of happiness, knowing that I was doing my work fairly well and at the same time 'going to school', even though it was only to evening classes twice a week.

Suddenly the scene changed. One evening I returned home from work and, upon opening the living–room door my father's eyes met mine; I saw and felt that something was amiss. He said nothing, but the atmosphere was charged with tension. My mother entered, but made no remark; I looked at one and then at the other and waited, for I suspected that there was going to be some complaint or other. I finished my meal and prepared to leave for night school, but my father arrested me with the words, 'Sit thysel' down, Siree, I want a word with thee. Thaiyt be fourteen in less than a couple of weeks time and I'm going to find thee a job in th' pit.'

I was dumbfounded and could not speak for a moment or two. 'The pit. . . a job in *the pit?*' I replied, not being able to understand such a change in the plans for my future. 'But I don't want to work in the pit, I'm doing alright as I am at the "Co–op".'

'Ah . . . so thou mightst be, but thou't getting only ten shillings a week and thy mother gives thee back eighteen pence of that for thy pocket. Things are difficult these days and thy mother needs the extra money which thou't get working down th' pit. Thy wages will be thirty five shillings a week and that's money compared with ten shillings; that's a consideration.'

I was stunned, but thirty five shillings a week sounded like a lot of money and I asked, 'Why do they pay a lad so much in the pit?'

'Thou mightst ask that lad. First of all you can't compare work in a shop with pit–work as thou will't find out for thyself. Wages have always been poor in the pit, but this war has made them realise the value of miners and coal. The demand for more coal sent up wages and reduced our hours to seven hours a shift. So, give thy notice in at th' "Co–op" and thou wut start work at th' pit as soon as thou art fourteen; lads can't go down th' pit before they are fourteen.'

Suddenly I remembered my holidays. 'But when I've been at the "Co–op" a year I am entitled to a few days holiday on full pay. Can't I wait till after that short holiday?' My father's eyes flashed and I knew that I had offended him.

'Thou has been towd what to do, so no more o' this, give in thy notice at th' week–end. There's a job for thee in th' pit.'

From experience I knew that it was useless even to plead with him or ask him to listen to reason; he was in no mood for that kind of approach. I was going to the pit, my parents had decided upon that and what I felt about it seemed not to matter at all. So I gave my notice at the 'Co–op' feeling very embarrassed as the Manager took in a deep breath and said, 'But what a pity, Harold, just as we had you trained to be a useful junior assistant. You deserve something better than pit–work.'

Others among the older assistants questioned the wisdom of my leaving a good safe job merely for the extra money each week. One said that economics were changing my course in life, but I knew that it was mere *poverty* which was taking me away from a job on the surface with fair prospects to a life in the dark bowels of the earth. I knew also that once established 'down the pit', it would be difficult for me ever

26

to become rehabilitated again into any other kind of occupation. These thoughts went round and round in my troubled brain until I locked up the front gate at the 'Co–op' for the last time, put on my jacket and went home.

2 First shift at the pit

'Harold . . . HAROLD . . .'

'Six o'clock . . . come on now.' The sound of my mother's voice calling my name awoke me from deep sleep. My mother, having had years of experience at calling miners at a very early hour, knew that sometimes it required more than one shout to bring response.

'Coming,' I mumbled and immediately shot out my leg into the cold air. We had been trained to do this from a very early age. I had recollections of my father telling us when we were very young that if miners do not respond at once when called, if they hesitate even for a second, there would be the risk of them dropping off to sleep again and being 'buzzed' at the pit and thus losing a day's work and pay. No miner, who remembered the lean years before the outbreak of the 1914 war took any risk of losing wages.

So, I quickly jumped out of bed, placed my feet on the cold linoleum and walked down the steep, narrow stairway. Behind the stairs' door hung our clothes. I hesitated as I remembered that very old clothes had been brought out for me from the 'old clothes box' and that they had been left overnight before the fire to become aired. I put these on with some resentment and, as I laced up my new clogs, I felt now that I was taking a step *down* in the word. Breakfast consisted of tea, toast and margarine. My mother moved around the living–room, cutting up my snapping and filling my water bottle, but she did not utter one word. She knew that I, the eldest of her three youngest children, was resenting having to work in the pit. Not one word was exchanged; everything there was to say on the subject of my working in the pit had been said and she knew that I was very sensitive to the fact that only the lack of money in our family was forcing me into this new life.

I rose from the table and put on my pit–coat, an old one belonging to one of my elder brothers and two sizes too large. I turned to my mother and said, 'Well, I'm going then, good morning.' I did not kiss her; no one kissed in our family. There was loyalty and obedience, there was respect, but no kissing, no outward display of affection. Life

28

was too grim, too hard for that kind of luxury. Kissing was for young lovers and for softies. My mother looked straight into my eyes and said softly, 'Good morning lad.' I was the fifth of our family to go down the pit, that is, the fifth including cousin Ralph; he was as much part of our family as any of us was. My father had known no other work since the age of twelve when he had first descended the older pit, the Number Six, better known as Big Pit. My eldest brother had gone into that same pit when he was old enough; the younger one (now lying dead in Flanders) had followed the same pattern. Ralph had never worked anywhere else but at the Fair Lady Pit at Leycett. So my mother, Margaret Brown, watched her fifth miner go 'down the entry' and off to the pit.

The iron tips of my new clogs rang through to the high ceiling of the entry. I knew that my mother was still standing and listening to the clatter of my clogs on the hard pavement; I knew it because I had watched her do this before when I had been obliged to be up at this early hour to get the washing–day moving immediately the men had gone to the pit. When I was young and had watched her standing there silently as the men left the house one by one, I used to wonder what she was thinking. The wives and mothers have their own individual manner of seeing their men off to the pit. Some follow their men right to the front door of the house and watch the men disappear; others say, 'Well, good morning then', as my mother had just said to me. Other women merely stand and wait for the sound of the clogs to disappear from earshot as my mother was doing, no doubt, as I turned from our entry and directed my steps from Abbey Street into Back Lane. Only the women themselves know the quality of the prayer they offer within their own breast for the safe return of their mining menfolk.

My snapping was wrapped in newspaper and I could feel it rubbing against my bare flesh, for I had placed the large bundle between my shirt and my body just as I had seen the others do on so many occasions. The bottle of water weighed heavily in the side–pocket of the too–large jacket and I was obliged to hold the bottle with my hand to prevent it hitting my knee at each step. I turned from Abbey Street into the quiet lane which skirts the southern side of Silverdale. In the same sense that a thousand ages in the sight of God are as an evening gone, so the recent events of the months since the visit of my headmaster to our house flashed through my mind. I walked up that lane as in a trance, oblivious of the trees, fields and hedgerows which had been my playmates during childhood and were still symbols of happy days. I re–lived every detail of those events; the headmaster's complimentary remarks, my work at Hambleton's shop and at the 'Co–op', and the enjoyment I had experienced at night school. A feeling of sickness came over me as I reflected upon the pleasure of sitting at my desk at night school and, quite audibly, I mumbled to myself, 'And what's the use of all that now?'

After a few seconds of mental vacuum I again addressed myself, 'Fancy, my second frustration in three years and I am still only fourteen.' My mind was too preoccupied for me to take much notice of things around me. Even when a night–shift worker greeted me on his way home I could only respond with a very unenthusiastic 'G' morning'. My apprehension increased and my fears mounted as the noises of the pit became louder and I recollected the stories which I had heard from my brothers about the heat and danger at Number Six, and how Ralph would not be outdone, declaring that there was no pit worse for heat than Leycett in the deep seams where he worked. Fear followed fear as each step took me nearer to the shaft which I should soon descend. 'What will it be like at the pit–bottom?' I asked myself.

I climbed the steep incline of the cleft between two high dirt–tips. There below me was the pit–bank with all its clangour—the loud crash of the cage as it was thrown down onto the supporting legs, heavy tools being dropped as colliers searched for their own shaft of newly–sharpened pick–blades at the blacksmith's shop, and the fierce hissing of steam escaping from bad joints in the pipes overhead. It all gave an impression of bedlam and confusion. Amid this strange uproar I tried to rehearse in my mind the instructions which my father had given to me on the previous evening.

I waited in a queue to reach the small window where my father was busy testing the lamps of his own men under his charge in Keele District. As I faced him he looked into my face for about two seconds, but to me it seemed like an age for I wondered what were his thoughts. I fancied that he was asking himself, 'Shall I make a pitman out of *this* one?' His countenance was non–committal as he told me to go over to get myself a lamp.

I waited in yet another queue and as I put my face into the small hole in the wall I was greeted with a smile. It was Mr Francis Durber the lampman. I knew him for he lived next door to our Sunday School and he was well–known in the village. He reached for a lamp without a word from me, then he checked that the lock was secure; with the same smile he remarked warmly, 'Making a start in th' pit then, Harold? I knew thou wast coming for thy faither towd me to have a lamp ready. This is a good lamp, Harold, number 672. Remember thy number and take care of thy light. Good luck lad.' I was very grateful for his warm smile and encouraging words for he had helped me over my first hurdle and I felt that I had at least one friend who was on my side. 'Thank you Mr Durber,' I said and I crossed to my father again to have the lamp tested. He said slowly and quite deliberately, 'Thy lamp is thy only light till thou hast finished thy shift. Dunna get thysel' in th' dark or thou't bey in trouble. Go down Keele dip and wait at th' bottom till thou art towd what to do; thaiyt be working with Les Bebbington in th' shunt.' And that was all he said. There were no instructions as to how I was to get to Keele dip, how far the shunt was from the pit–bottom and what, if any, were the hazards. Within me, I knew that had I dared to ask *why* I had not been given more information regarding directions, I should have received the usual reply, so often heard in our family, 'Well, hastna got a tongue in thy head then?'

'Go down Keele dip and wait.' That was the instruction and I stood still for a moment repeating it. I turned towards the two shafts, Number 14 and Number 15. There was a queue at each one so I joined my third queue since coming on to the pit–bank. As we shuffled along nearer to the cage each time it was re–loaded with its human cargo, I examined my Davy lamp and gazed at the tiny flame. I tried to understand the mystery of its 'safety' and tried also to recall the lesson we had received on the Davy lamp at our Elementary School. It was still dark and I remember feeling the fresh, clean air blowing on to my face as it came in over the flat Cheshire Plain from the sea. Soon I was near enough to the shaft to see the cage suddenly appear with its four heavy chains drawn together to the cap which held it securely to the pit–rope. As the cage came to the level of the pit–bank it lifted the great heavy safety gate which surrounded the shaft when the cage was in motion. My eyes followed the cap and the chains as they were lifted high into the air. The banksman threw the lever forward and the cage was dropped with a bump on to the legs. The black–faced night men poured out first from one deck, then the next. Gently the cage was

31

lifted from the legs, and a signal came from the pit–bottom that the cage in the other shaft was loaded and ready. DOWN it went, like a heavy stone. The shaking of the rope indicated the strain of the heavy cage and its cargo.

I had to admire the quick action of the banksman and his reaction to signals from pit–bottom. I was yet to learn how vital was the speed of this operation of 'decking' if all men were to be transported down the shaft so that they could get to their places of work at the far ends of their districts by seven o'clock. Already I had estimated the number of hours I should be 'down there' before I should be able to stand upright again and take in good clean air. I knew a lot about pit–life already, for our community depended upon pits for its existence and I had known little else but pit–language since the days that I could understand ordinary conversation. So many mixed thoughts and fears as I waited for the next appearance of the cage for this next one was 'mine'. I touched my snapping and water bottle to make sure that they were there and I again looked at the flame of my lamp to re–assure myself that I did have a light. Suddenly, there it was—the cage. I had already counted the men in front of me and thanked God that I should be riding on the top deck. Had there been an observer on hand he would without doubt have seen in me every sign of a very frightened and nervous little boy. The night men tumbled out of the cage and I was almost carried forward by the weight of the queue behind me.

In my anxiety not to lose my light I looked down to see where I was putting my feet. I gave my shin a sharp bang on the rails which run through the cage, but the thick leather of my new clogs protected my flesh. I fell, head–first, into the body of the man immediately in front of me who had turned about to be facing the entrance ready for getting off the cage at the bottom. His dirty pit–clothes smelled of stale sweat and some other substance which cannot be described, but which all miners know belongs 'only to the pit'. The smell of the bodies packed like sardines, the smell of the cage itself, plus the fact that, being small, my face was pressed tightly into the belly of a fully grown man, already nauseated me. We were jerked up a few inches and I knew that the supporting legs had been withdrawn. I was very afraid and when I have heard young pit–men boast that they were not afraid when they hung on the end of a pit–rope for the first time, I have always disputed it. My mind was now on the gaping hole beneath me, some four

32

hundred yards of complete darkness with nothing but the strength of the pit–rope to save us from a terrible death. Suddenly we were hurtled downwards and above my head there was a terrific . . . C–R–A–S–H. I gasped and shouted, 'Oh! . . Oh! . .'

'It's owraight . . . it's owraight . . . little mon . . . it's only th' gate.' A kind old miner had put his arm around my shoulders as he quickly realised that this was my first ride down the shaft. He soon assured me by his soft words and firm embrace with his strong, rough hands that there was nothing to fear. I became conscious of a slackening in speed and I knew that we should soon be at pit–bottom.

'Someone should a' towd thee about the' gate, 'said the man who still held his hands on my shoulders. He said it with a warm tone in his voice and I was grateful to him.

Another voice from the front of the cage called out, 'Thy faither should a' towd thee about that bloody gate, it inna raight.'

The older man spoke again as the cage slowly dropped on to the supporting legs, 'Who is th' lad then?'

'Whey, dostna know?' answered the other. 'He's George Brown's lad. This is his fost day in th' pit. I heard Durber telling him to look after his lamp.'

Suddenly there was light; it was a strange experience to emerge from the dark, cold shaft into the pit–bottom, so far down in the earth, all lit up with electricity. This was something I hardly expected. As I walked off the cage I became aware of a most appalling stench, a horrible obnoxious smell. It had been bad enough on the cage tightly–packed among the men with their dirty clothes soaked in old, stale sweat; even that had made me feel sick, but now, this stink! I wanted to vomit. I did not know that it was going to take many weeks for my nostrils to become accustomed to this dreadful stench.

The older man who had befriended me on the cage paused as I, not knowing which direction to take, stood gazing around me. I was bewildered since there was a road to my right leading off into some district and before me there was the main shunt of the pit–bottom all alight. The onsetter in charge of the cage was much too busy with his work to be bothered with me. Every miner leaving the cage proceeded hurriedly up through this shunt and into his own district where he worked. But my friend on the cage waited and, watching me, after a moment he said, 'Where art working lad?'

'Down Keele dip, mister. I'm working in the shunt with Les

Bebbington,' I answered him.

The man exchanged glances with the onsetter. 'Oh . . . down in Keele art working? Well, come along wi' maiy, I'll show thee Keele dip. I work in South East so I conna but show thee where Keele is.'

How grateful I felt toward this older miner; I had yet to learn that this man's conduct was typical among miners. I had heard during my growing—up years of this bond among men underground. I had heard it said in our house that hazard and danger were the common lot among miners and that all barriers come down once the men are underground. Here, in my first few minutes after stepping off the cage for the first time, I was experiencing that very thing. I walked with careful steps behind the man who already, I felt, had become my friend. I was too apprehensive in mind to observe the great pumps and engines which were housed in the pit—bottom. After a hundred and fifty yards or so, the road divided and the electric light came to an end. There were now three roads leading to different districts.

The old man put down his shaft of pick—blades and took my arm. 'Here thou art lad, here is Keele dip'. 'That is South East where I work and just along there up that smaller road is Spencroft. Nagh, watch thy step as thou goest down Keele for it's very steer. Hast got thyself a lamp—shade?'

'What's a lamp—shade, mister?' I asked timidly.

'Whey lad, if thou dostna have a lamp—shade, thy light will blind th' men travelling behind thee and they will kick thy light out to teach thee a lesson and then thou't be in th' dark. Here thou *must* have a shade at th' back of thy lamp. I'll make thee one out of this bit of brown paper which I always carry with me. But thou must get thyself a proper shade or thou't be in trouble with thy mates. Oh . . . another thing, give thyself at least five minutes to get thy eyesight before entering the dark dip. It's over three quarters of a mile to the bottom of the main dip.' This generous man tore off a piece of his precious brown paper and wove it among the metal bars which protect the glass of the Davy lamp; he then turned and walked into the darkness of South East district.

'Thank you, mister, very much,' I shouted as he moved away.

'It's owraight, lad, just keep thy head down in that dip or thou't bash thy brains out on th' low girders.'

I hesitated for a moment feeling warm inside on account of this comforting experience with this older miner. I raised my lamp and examined the lamp–shade which he had made for me and realised just how important a bit of brown paper can become. I walked over to the entrance of the dark dip itself, but I had not yet got my eyesight and I put my foot into a deep sleeper–hole; I almost fell headlong, but fortunately managed to remain upright and save my light. Black–faced miners walked towards the pit while the clean–faced ones hurried to their work.

After only a few minutes three young men walked briskly up to the spot where I stood. As they came to a halt and took off their coats they displayed a jovial air and joked among themselves; they appeared to be very happy. Suddenly one shouted out, 'Eh up, Eh up . . . it's Harold Brown. Damn me if he isna working in th' pit. Ah–do Harold, this thy fost day then?'

'Ah–do Jack. Yes it is my fost day. Dost thaiy work in Keele then?'

'Yes I do, 'ere come into this man–hole and put thy jacket in th' corner along wi' mine. Thou costna tak' thy jacket down th' dip, thou't bey roasted alive by when thou getst to Plate Landing. Keep thy snappin' in thy shirt and carry thy water bottle in thy hand. Thou't have to get thysel' one o' these bottles which fits to thy strap, and then thou't have one hand free for thy dip–stick; hastna got a dip–stick then?'

'No–one told me that I should need a dip–stick.'

'Well, 'arold, they dunna all have one, but it helps thee at th' end of th' shift when thou are running up th' dip at Loosit [end of the shift]. Hast got thy eyesight yet?'

'Yes, I think so. I can see things better now.'

It was Jack Wilson who was advising me and trying his best to befriend me. The other two had shown only passing interest in me. All three waited for a few moments to get their eyes accustomed to the darkness. One of these other two men was Ambrose Rowley whom I knew; the third I knew only by name, he was Redmond D'Arcy. Redmond was one of several brothers who were all well–known as good workers. All three seemed to be enjoying the novelty of having a newcomer to the pit among them.

'Ready?', shouted Redmond (he was hooker–up at the bottom of Keele dip) and at once we all four set off down the steep dip. Redmond carried with him an air of slight superiority for he had a responsible job. He had a silent manner; I had heard my father talk of this man and I gathered that he commanded some respect from the officials not only because he did his work well, but because he was able to conduct himself without the use of foul language and bawdy stories. He had a keen sense of fairplay which he had from time to time expressed, even to higher officials.

As Redmond D'Arcy had indicated that he was on his way into the dip and that he expected the others to follow, Johnny Wilson said to me with a warm voice, 'Come on Harold, just keep thysel' behind maiy and when I shout "LOW" make sure that thou getst thy head well down for there's a lot of bloody low girders in this dip and thaiyt bey knocked sick if thaiy walkst into one. When thou canst stand upright again I'll shout "HIGH". Hast got thy lamp–shade fixed owraight?' He paused and took a glance to his rear; I raised my lamp for him to make sure everything was correct. When I called him 'Jack' he reminded me that he was always known as Johnny; he swore like a trooper, but he was a likeable young man and already I was grateful to him. I think I would have been grateful to the Devil himself if he'd given me the help and advice I was receiving from these young men. I was amazed at the speed with which they walked down this steep incline with only enough headroom to take a pit waggon, some 4 ft 6 in high. It was Redmond D'Arcy who passed the instructions back as to the height of the roof and occasionally Johnny shouted back to me, 'Deep sleeper–hole, mind thy step 'arold.' From time to time we met night–shift workers going out who seemed only too pleased to pause and let us pass if only for a second or two. There were several flat stretches in the dip for which I particularly was grateful. At the Steer it was necessary to proceed very slowly for one false step and a man could go hurtling forward and lose control of his step. My back was breaking for it was unaccustomed to this back–bending operation. I banged my head twice, but I dared not complain or show any sign of weakness for I had received more than one lecture at home on the standard expected not only from miners in general but from Browns in particular. Indeed, I still had recollections of my two elder brothers being so lectured on this very subject when I was only just old enough to understand.

At last we were at the bottom of the dip. The roof was low, so low in fact that the waggons standing there almost touched it. The road continued straight down, but I could see that the main road turned sharply to the left; this wider road to the left, I was told, led to the coal face and the far end. Redmond D'Arcy spoke to me for the first time.

There was no North Staffordshire colloquialism in his manner of speech, no 'twang' as the Staffordshire people refer to it. Just a quiet, direct way of expression. He looked at me and smiled, lifting his lamp to my face, 'Your father told me that you would be with us today, Harold. You will work with Les Bebbington, do you know him?'

'Know him? I should think I do. He was in my class at school and for a twelve–month we sat at the same desk.'

'Good, that will make things easier. Just remain here till he arrives. He is nearly always one of the last down; he waits to come along with the Road–doggy.

As the other two men rested by going down on to one knee I thought of Johnny Wilson's remarks about a dip–stick to help in the 'running' up the dip at the end of the shift. As I tried to accustom myself to my new surroundings the words kept drumming through my mind, 'Running . . . *running,* up that incline, how is it possible?' What a new world this was to me. Perspiration was running from every pore in my body. The physical effort of the journey down the dip had taxed me to the limit and, already, I felt exhausted. Gone was the flow of cool air which I had felt at the pit–bottom and now there was a dull stillness in the atmosphere. Although the shift had not commenced, the smell of sweaty bodies pervaded the confined space where we all rested. Already I could feel the heat of which I had heard so often from my brothers and I was yet a good distance from the coalface itself. But I was fascinated with the roof; it was very low indeed, but it was smooth and even, with no rough places protruding, nothing to cause injury. The even flatness covered the whole of this dip–bottom and, although we could not stand upright, it was solid and smooth; to me it was a safe place. I sat gazing at this roof with a questioning mind for I could not understand why the roof should be so even and solid at this place.

Redmond D'Arcy was already at work passing his hands over the couplings between the journey of loads which stood there. He then passed his hands over the top of the coal, making sure that no large lumps protruded above the top of the waggon. Men were now almost running down the dip in order to get to the face by the stroke of seven o'clock. I became curious about a glimmer of light down below the bottom of the dip; I could also hear the throb of some sort of engine.

'What's down there?' I kept asking myself.

Soon there was the noise of voices; it seemed as though an army was approaching. Suddenly it was all bustle and hurry; all noise and fuss, and yet only two men had been added to our company. The one I recognised was my old schoolboy friend, Les Bebbington, who hurried into the shunt. The other man of the two made his presence known by shouting out an order, for he was obviously the one in charge.

'Come on, Redmond, bloody well knock him in, it's two minutes past thou knowst.' Redmond made no remark, but obeyed the order and signalled to the engineman in pit–bottom that the dip was clear of men and that the journey of empties could be turned into the dip. The sound of the bell fascinated me—communication by means of bell–signals, all that distance from the shaft to this junction in coal–traffic, nearly a mile away. No–one took the slightest notice of me, although I did detect that the man in authority with the loud voice gave me a quick glance. I began to wonder 'Is he ignoring me on purpose? Why does he not tell me what to do?' But I was quite wrong for at that moment the bell rang again, ONE TWO . . . ONE TWO—the signal, two twos. The engineman was now announcing that his journey was approaching dip–bottom and that he required the same signal back again, giving him permission to run the journey into the dip–bottom. The engineman's signal was returned, ONE TWO . . . ONE TWO.

In seconds there was the roar of the ten empty waggons as they thundered down into dip–bottom. Redmond D'Arcy stood at the side of the track, body bent on account of the very low roof, but he assumed the attitude of a crouching ape, swaying his body to make it coincide with the rhythm and speed of the moving waggons. As the last waggon became level with him, Redmond made a great leap on to the rope, pushing down his weight to create enough slack to enable him to withdraw the pin from the shackle and disengage the rope from the journey. He maintained his balance by holding on to the top of the wagon with his right hand although there was only just enough space for his fingers. Again he made another leap, this time across to the first load on the train of loaded waggons and, even before he had the shackle–pin firmly fixed he bellowed, 'Knock him up.' The man in charge stood with his hand already on the bell wire and I heard the steady signal . . . ONE . . . TWO . . . THREE. The rope slowly began to tighten and the journey of loads moved away. Redmond grabbed his lamp from a hook on a nearby post, dropped down on to one knee and examined each coupling as the journey gathered speed.

I was fascinated by the language of both men and bell–signals. The expert manner with which Redmond D'Arcy had crouched down, leapt on to the rope and jumped over to the loads with such precision, all gave me pleasure, so much so, that I almost forgot that I was so very deep down in the crust of the earth. I had gone down, down deep through the shaft, then further down, down, down deeper still as I had descended Keele dip.

Once the journey was out of earshot, the man in charge came over to me; he lowered himself on to his knee and with slight superciliousness remarked, 'So, thou art Harold Brown then? . . . Hmmmmm! . . . George Brown's lad eh!?'

A strange feeling gave me the impression that this man did not like my father, but I could not be sure. 'Come on, lad,' he shouted with a superior air. 'Dunna just idle there, I'll show thee what thou hast to do. My name's Arthur Ricket and I'm th' Road–doggy here. Thaiyt workin' in this shunt with Les. He will show thee what thou hast to do. But I dunna want to hear of any sky–larking or any other trouble between you two lads. Understand nagh?'

He had offended me as he had shouted to me—'Come on, lad, dunna just idle there.' The word 'IDLE' did not belong to the Brown family and I felt that he had made an unfair thrust. However, I walked further into the shunt where Les Bebbington, who was stripped to the waist, was hard at work. I saw what had to be done and soon my shirt was off and I was glad of it for sweat oozed from my every pore. Les came to my side and said breathlessly, 'Ah–do Harold.' Nothing more. He assumed that common gumption would tell me what the work was and that I should automatically get on with it. My mind went back to the shop and silently I said to myself, 'This is going to be different to that kind of work.'

My first waggon dropped off the road. Les jumped to my assistance. 'Owd on 'arold,' he said showing that he was pleased to know more about the job that I did. Owd on a bit, I should a' towd thee. Each time thy waggon goes over this particular joint in th' rails, thou hast to give the arse–end of thy waggon a sharp twist. 'Ere . . . I'll show thee what I mean.' I stood aside. Les took my waggon and, by a trick I was yet to learn, heaved it back onto the rails. He then gave me a demonstration in the art of twisting the arse–end of my waggon over this joint in the rails. Here was my first lesson in

39

waggoning in a very confined space for there was hardly room for a small boy to stand between the row of waggons and the row of loads on the opposite track. But there was *headroom* and this was luxury. The traffic in the shunt demanded the facility to stand upright. The good roof had been sacrificed here in the shunt for I could see that it had recently been ripped. But the thing which impressed me about my first lesson in pit–work was that it came from Les who I always regarded as being just a step below me in our classroom work at school. He was a likeable lad and we had always been pals; it was good fortune for me that I had been put with him on my first day.

This pushing up of empty waggons and bringing down of loads went on without stop and my body was soon aching from head to toe. Some waggons which had not been greased at the axles took all my strength and often it required both of us to deal with them. Waggons, loads, sleepers, rails, joints and very deep holes between the sleepers, these were now all my world plus the handicap of almost complete darkness in some parts of the shunt. Our two lamps were placed so as to give us light where it was most needed; otherwise we fumbled about in semi-darkness and relied upon our ability to feel our way about. I could hear the bell-signals at the bottom of the dip and derived pleasure from the fact that I already had learned something of their signifance, but I was too engrossed in my new slave–like work to indulge in any romantic ideas of my being involved in the major trafic complex at the bottom of Keele dip. There was no time for conversation for the waggons were always there to be pushed up and the loads came down from the waggoning lengths above us in a constant flow.

At long last we heard the bell–signal from the pit–bottom, ONE . . . TWO . . . THREE . . . FOUR . . . FIVE. 'Thank the Lord for that,' shouted Les. 'Harold it's snappin' time. Oh, where hast put thy snappin'?' I pointed to a small ledge, but Les snatched it away and put it up on a girder, the highest point he could find. 'Look thee, Harold, always put it as high as possible. There are mice and cockroaches in th' pit, they dunna get up that high easily!' I took down my snapping again and we threw ourselves down on to two slabs of timber for a few minutes before we began to eat. Our bodies were exhausted.

'Mice . . . cockroaches . . . what next?' I thought as I lay there. Les warned me of the vermin and said, 'Harold, always open each piece of snappin' before thou hast a bite; mice–dirt or cockroach dunna go well with jam or apple sandwiches thou knowest.' I thanked him and we laughed together once the sweat had been wiped away from our faces with the tails of our shirts. I could hardly believe that I was now relying upon Les Bebbington for advice and guidance and that I was feeling so utterly grateful to him. We ate our snapping in silence and the twenty minutes was soon over.

'Before I forget,' Les said, as he rose from his plank, 'Try to persuade thy mother to get thee a proper snappin'—tin to keep the little devils out of thy food.'

We walked to the top of the shunt to look at two prostrate bodies lying flat on the ground. Their bodies and faces were completely black and I did not recognise them. Les informed me that I knew them alright. They were Johnny Thacker out of Brooke Street and George Kitchen from Newcastle.

'I've heard a lot of swearing from these two, Les, during this first half of the shift.'

Les laughed loudly, 'Cursing, swearing dost say Harold? Bugger me, thou't curse and swear I can tell thee if ever thy faither puts thee on that waggoning length. Its a hard long stretch o' road to waggon on I can tell thee.'

We heard the rumble of waggons at the bottom of the shunt and back we went to our slave—like duties. Les took my arm, 'Here 'arold, thou must have a go at lockering loads now. Thou't have to learn how to locker thou knowst.'

He demonstrated how to stand with an iron bar in one hand and throw it into the wheels of the loads as they ran down into the shunt. He looked worried as I missed the first load, but he was below me with his locker at the ready. He said calmly, 'Harold, we have to make sure that loads coming down too fast dunna crash down on those standing in th' shunt or they become cross—buffered. Then there's more bloody hard work AND a damn good cursing from th' Road—doggy. To make sure that thou dostna get a pile—up of loads through thy mis—locking, never take thy first locker out until thou hast thy second locker firmly in the load behind. Dost understand maiy nagh?'

'Yes, Les, I think I understand what has to be done.'

And so I learned yet another element of pit—work from Les. The heavy mauling and groaning went on and I began to suspect that there were ninety minutes to the hour down in these dark tunnels under the earth. All the warnings I had heard in childhood about keeping away from pits came rushing into my mind. I remembered one miner telling his son who was my age . . . 'When you see the pulley–wheels going round, run as fast as you can away from them.' Another common saying was 'ANYTHING IS BETTER THAN PITWORK.' I was now realising the truth of all that.

At long last came the signal at dip–bottom. ONE . . . TWO . . . THREE. . . FOUR . . . FIVE . . . SIX . . . SEVEN. 'LOOSIT', shouted Redmond D'Arcy at the top of his voice. Suddenly there was not a light to be seen at dip–bottom and all was in deadly silence. As if spirited away, the three men and the Road–doggy had made a rush for the pit.

'How can they run?' I again questioned in my mind. 'Running up that steep dip at the end of a heavy day.'

I felt weary as I put on my shirt and said to Les, 'Dunna wait for me, Les, I shanna be running up that dip.'

'Oh Harold, I dunna run up. Everybody makes a mad rush for th' cage, but I take it easy like. Come on, we'll wander up quietly together. Soon the workers from he waggoning lengths, then the loaders and colliers from the coalface began to overtake us. We stood aside to let them pass and I wondered how they found the energy to rush up the incline.

Les stopped, turned to me and asked, 'How dost feel Harold, art stiff?'

'I can't describe how I feel, Les. Tired out and I ache in every part of my frame.'

'Oh, thou't bey owraight after thy second shift. Thaiyt bey stiff owraight tomorrow morning, but after another shift in that shunt all thy stiffness will be gone.'

Les's quiet, deliberate manner was exactly as it had been during school days. I knew that he had bad eyes and I suspected that this was responsible for his retiring ways.

The sight of the electric lights at the pit–bottom was welcome. There was a queue waiting for the cage as we squeezed our way forward between waggons and loads. Les pointed up into the roof to Keele dip engine and nearby on ground level was the engine for South East district. Near the shaft we saw a huge long pump–house with the electric pumps screaming in high pitch as they pumped the water to the surface.

A man with a club–foot attended the pumps and Les remarked, 'See that chap there looking after the pumps, well that's Harry Wilson. He is brother to Johnny Wilson who brought thee down th' dip this morning.'

'What about this awful stink Les. What the hell is it?'

Les laughed and looked straight into my face. 'Thaiyt get used to it in time Harold. You see, Harold, it takes thy guts a long time to get properly used to it. It is all the muck that runs down the side of the shaft and through the earth with the water. Most of it finds its way into th' sump and I dunna suppose it ever gets cleaned out. The water running down through the earth brings with it various chemicals I suppose, thou knowst; lime, iron, and God knows what else. It settles in th' sump and just lies there, stinking the bloody place out.'

'Harold,' said Les, pausing and adopting the attitude of one about to give a lecture. 'Dost 'ear . . . when the cage gets to th' top, it will lift the safety–gate up with it and if he is winding a bit quick like it will give a heavy thud above thy head as the cage hits it. So, dunna let it frighten thee. It inna so bad going up but when th' cage drops like a ten–ton stone down into th' shaft and th' gate crashes down above they head it can frighten the damn life out o' thee if thou art not expecting it. Did anyone warn thee this morning, Harold?'

'No, Les, no–one mentioned that gate and it frightened the wits out o' maiy I can tell thee. I shall tell my faither about that gate.'

'I shouldna, I shouldna, Harold. Thou knowst what he is. These owd miners think it weakness if any of the young ones complain. Dunna give th' game away that thou hast been at all frightened; take it in thy stride; thy faither might think thy fear may injure his reputation; he's got one, thou knowst that.'

Again we were packed like sardines in the cage and 'swish' we went up the shaft through the cold air. All the disadvantages of the pit disappeared from my mind as we suddenly burst into daylight. What an experience that was. DAYLIGHT after eight hours of utter darkness. I had never imagined that daylight, which we had always taken for granted, could be such a comfort and so reassuring. It was too bright for me and I had to shield my eyes for a few seconds. As we walked home together I asked, 'Les, why haven't we seen my father today? He *is* in charge of Keele isn't he?'

'Ah, he is, but he dunna always come through th' shunt. If he has a lot of shots to fire, he conna get away from th' face.' I marvelled at how

43

much Les had learned in the few weeks he had been in the pit. Arriving home I noticed again how my new clogs rang through our entry, the high empty space echoing every sound.

Mrs Brown, our neighbour opened her door at the top of the entry and said, 'My . . . My!' We now have another *man* in our entry.' Mrs Mary Anne Brown was our next–door neighbour, for there were two Browns in the same entry, and a more genial person was never born. 'Another *man* in the entry,' Mrs Brown had said. It is true, work down the pit *is* a man's life; the weakling could not survive. Having achieved this survival by working my first shift without mishap I felt that I had earned Mrs Brown's remark.

'No', I said under my breath as I unlatched our top entry–door, 'No, I must not mention that gate to my father.'

I followed the traditional procedure of miners. I took off my clogs and jacket, washed my hands and entered the living–room to eat my meal which had already been placed on the table. But the bright fire, the home–made rug before it, and the snug cosiness of the room were all too much for my tired–out body. I had not the strength to sit up and eat my food. I threw myself down on the rug before the fire and in a moment was lost in sleep. Sleep was more vital than food at that moment and my mother let me sleep on for she knew how I felt; she had watched her other sons return from the pit exhausted after their first shift. When at length I was roused my father had already finished his meal and my favourite dish had been kept hot for me. Meat and potato pie; a *whole* pie just for myself. My father said nothing, but he smiled as he watched me do justice to my mother's cooking, every morsel of the pie devoured. I knew the routine without having to be told. I retired to the wash–house where I 'got my delf–dirt off'. Then into cleaner clothes and the rocking–chair in 'the parlour'. I was weary in body for the heavy work had taken a toll; in some respects I was not easy in mind, but as I put my head back to rest I found myself recapitulating the events of the day.

A warmth crept over me as I began to realise that the dominant features of the day were not the dirt, the cage, the fear, the stink; no! what stood to the fore in my thoughts was the help I had received with such cheerful spirit from men who need not have noticed me. The older miner who had befriended me on the cage; Johnny Wilson making sure that my lamp–shade was properly adjusted and that I came to no harm on the hazardous journey down the steep dip; the reassuring words of Redmond D'Arcy . . . 'It is heavy at first, but you'll get used to it in time.' Then there was Les Bebbington. Although just a pit–boy, not long having started in the pit, he had done his best to keep me from harm and make my first day underground as cheerful as possible. All these constituents of my first day down the pit unwound in my mind as I sat gently rocking to and fro in the chair; I was trying hard to adjust my mind to this new situation in my life. The comradeship which I had experienced on this first day was something new to me. The inner warmth which the memory of it all brought made me begin to lose the dread of walking up Back Lane to the pit next day. Indeed, I found it difficult to settle to my usual evening's work. In any case, there was now no necessity for night school and, with shorter hours at the pit, I should have more time for my study at the piano.

My father, who had worked in the pit since the age of twelve, had become qualified as an official at an early age. I did not remember him as anything else but as a fireman or a deputy as such officials are sometimes called. He had spent all his working–life in the old Number Six Pit where working conditions were much more difficult than they were at the pit where we now both worked. Number Six had closed and my father was taken on at once as an official at Kents Lane Pit. He and others from the Number Six often talked of the comparative comfort of working on the gentler gradients at Kents Lane. Any suggestion of complaint would bring the warning, 'Mind you don't find yourself in conditions like those in the old Number Six.'

My father was a man of few words as far as his work was concerned; a favourite expression was, when he was offering advice to us young ones, 'A silent tongue is a wise head.' 'Silence is Golden,' he would say sometimes.

As I sat enjoying the relaxation in the rocking chair after the physical demands which had been made upon my young frame, my father entered the room. Immediately, I offered him the chair for he spent hours in it with the newspaper. He walked over to the window and indicated that he did not wish to sit. From time to time he glanced at me but I was in doubt as to the quality of his glances and his silence. I knew better than to disturb such a situation by trying to establish conversation. He took another chair and held the daily paper as if to read, but I noticed that his eyes were on the ceiling. I knew that something was coming and after five minutes or so of this silence he said casually, 'Harold, I've got something to say to thee. There are things thaiyt have to hear.' My spine went cold, I felt uneasy and I wondered what I had done wrong on my first day in his district down the pit. I was quite oblivious of the importance of the minutes which were now about to pass. I looked at him, then reached up to the bookshelf as if to indicate that I wanted to do some study. I could hear my mother busy in the next room and wondered if *she* knew what was in store for me. My father raised his voice, 'Thaiyt 'ave leave that alone for half an hour till I've finished with thee.' My fear increased still further as he still hesitated, said nothing, but just gazed into my eyes.

As if searching for words he said very slowly, 'My lad, your life in the pit started today. We Browns have always been miners, at least the last three generations have been. You will find life very hard in the pit, but it will make a man of you; you will find fine characters among miners—men who work hard for their wages and observe the pit regulations. Such men are brave and courteous, they are second to none among working men.'

Here, I felt an urge to interrupt and tell him how I had already observed that very thing, but I took in a deep breath and arrested my impulse for already I was beginning to feel a glow of pride within my inner self.

'You will also find a few lazy people, but there are not many among miners; these people do not fit in with the standards and demands of pit–work. As an official I have to deal with laziness sometimes. You will hear things said about me by this kind of person which is not always complimentary because I insist on all safety rules and regulations being carried out to the letter. Now, Harold, listen very carefully to what I am going to say to you. You will make me very proud if anyone relates to me that you are a good, hard worker, a *real* Brown. *But,* if I hear that you are lazy, if you do not maintain the tradition for conscientious work which has always been associated with the Browns, then you can pack your bag and leave this house for I could not tolerate the humiliation of having a lazy one in our family. If ever you are in doubt as to whether you have done your full day's work, there'll be no harm in you doing the extra bit to make sure of it.'

I took in another deep breath and nodded to indicate that I had understood, but the responsibility of being a Brown was beginning to dawn upon me, with all its implications.

My father again went on, 'Never let it come to my ears that you have been shirking your work. There are several branches of Browns in this district and in each case, it is the same story, *"The Browns are workers".'*

He fell into silence again. I gazed at his face and admired him. There he was, the mining official, the strict father, the proud Staffordshire man. I shared the silence with him and my eyes fell to the small well–worn rug before the fireplace. As my eyes traced the edge of what little of the pattern there was left, I felt that my father was now looking at me and trying to estimate how far I had understood just what he had been trying to inculcate into me. I was not sure how to deal with this situation, so I remained silent.

My father came over to my side of the room, I had now risen from the chair and we stood together side by side. He placed his hand on my shoulder and said firmly, 'So I shall rely upon thee to maintain and uphold this reputation of ours. Now, if you prove yourself worthy of the name Brown you will never have to fear any man. We have never owed a penny to anyone; what we have we pay for, what we cannot pay for we go without. We have seen some very hard times, but we have survived. You will not find life a bed of roses, but if you are a true Brown you will be able to hold your head high and be proud. No man will be better than you are and you must never, never allow anyone to tread you down. Remember now, your family honour and reputation in the pit will help you to preserve your dignity and self respect. I am no scholar, but I did once hear someone use a few Latin words which meant *Always the Truth.* Stick to the truth lad and thou't not go far wrong. So I pass these things on to you and perhaps you, in turn, will pass the same words on to your children. My own father drilled it into me that speaking the truth and being kind to old people was a golden rule for life. Do not let words flow idly from your lips, especially when you are speaking about people. Guard your words and God bless you in all you do.'

Again there was silence. Hardly knowing how to reply to a speech of this quality I said softly, 'I'll do my best, father, not to let you down.'

I stood there as if struck. My body was aching from my day of heavy toil. I was bewildered in mind by the appalling conditions I had found in the pit and, although I was warmed and cheered by the

47

comradeship which I had experienced, the harshness of pit–life for the rest of my working days was not a very cheerful prospect. And now, *this*. This awful threat of expulsion from my family and home if I failed to maintain the reputation of the Browns as a worker. I felt the responsibility of being a Brown.

Forsaking any idea of study or piano practice that evening I walked down the entry and into Back Lane; I let my steps take me which direction they pleased, but my pace was not that of a young boy of fourteen for my mind was very deeply concerned. In the darkness I came to Rosemary Field where it was possible to walk up a narrow track to the high road which skirted Keele Park. There I sat gazing at the ocean of twinkling lights of Newcastle–under–Lyme and the Potteries towns. I searched my mind for a solution, some means by which I could extract myself from the trap I was in, the trap of working down the pit for the rest of my life. Peace of mind did not come, nor enlightenment. It seemed useless now for me to nurture in my breast any longer the ambition to enjoy some form of better education. I turned my steps homeward with a heavy heart, but for some reason I did not go straight home, turning right at the old Racecourse Pit into Downing Street, making a detour merely for the walk. As I came into the High Street I saw the lights on in the school. I stopped suddenly and took in a sharp breath.

'That's it, that's it,' I almost shouted out loudly. 'I'll go to night school several nights a week and try and make up for what I've lost.' My mind became lighter; I went home and straight to bed. My tired body got the better of my active mind and soon I was asleep.

The sound of my father closing the entry–door awoke me even before my mother called to say that it was six o'clock. Reaching the bottom of the stairs, there stood my mother about to call me.

'Oh, so thou't awake then. Couldstna sleep then?' she enquired.

'Oh yes, I slept alright, but I heard father go out.'

There was an element of resolve in my every movement. The cold water refreshed me and after my breakfast I walked briskly from the house as though leaving on some urgent mission. There was now no escape; I was involved in the occupation of 'working down the pit'. Within my mind I repeated several times, 'Better make the best of it and try to become a good pitman.'

I tried to ignore the stiffness in my legs, arms and shoulders. Each step up the lane seemed to give my whole body a jolt, but I was now determined to push my aches aside; after all, I had been stiff before. Yesterday, on my first day, my mind was too preoccupied for me to notice much as I made my way to the pit. But today was different, for since my father's lecture and my quiet reverie at the top of Rosemary Field I had come to terms with my new status in life. This second morning I was noticing things. Several night–shift workers exchanged the usual greeting, 'Ah–do Harold, made a start in th' pit then?'

I looked to my right and there I observed the girls at work in the fustian mill, walking back and forth, back and forth, in the light of their flickering candles. I actually stopped and watched them for a moment, feeling sorry that *they* had to start work at six o'clock. As they turned at the end of their long lengths of cloths, reinserted their fine needle into the cloth and started back on the endless walk, I recollected that they worked on and on like that every working–day of their lives for ten shillings a week. Over sixty hours a week and they had to provide their own candles from the ten shillings wages. I could not understand why I should suddenly stop on my way to the pit, watch those girls hard at work, and feel so sorry for them. Perhaps it was my realisation that now I, too, was among that army of people who were having to 'slave' to obtain enough money merely to survive.

A few yards further on, opposite the Sneyd Arms public house, better known locally as 'The Bush', there was Les Bebbington waiting.

''Morning Les. This inna thy road up to th' pit. What art doing here?'

He did not reply at once, but joined me in my walk. At length he said with a chuckle, 'No, it inna my road, I usually go through Brook Street and up to th' pit past the station. I thought I'd come a bit early this morning, catch thee and give thee a few tips like. Thou knowst, show thee th' ropes a bit. I know a good place at th' top of th' dip for thy jacket; thou wanst a bit of instruction about going down the dip thou knowst. When I fost started, I bashed my head something cruel on those low girders because nobody towd maiy about them, that they were so low.'

Again, I felt warmed that my old schoolboy friend was going to work a bit early just to help me. I eased my pace and said, 'Well Les, thanks. It is good of thee.'

'Oh, dunna expect it after today. Thaiyt 'avin' only one day of my tuition thou knowst. By the way, art stiff Harold?'

'Stiff, *stiff*, dost say Les? Every time I put my foot down on this hard road, I get a jolt up my spine and every muscle in my body feels like a plank!'

Les laughed and touched my arm, 'Thaiyt bey owraight at the end o' this shift, thou canst depend upon that Harold.'

Down the slope we marched together; Les pushed me forward to the hole in the lamphouse and proudly I shouted 'Sixseventwo.' Francis Durber handed my lamp to me without comment for I was now 'one of them'.

3 The work of two boys

My father tested my lamp, but said nothing. He looked straight into my eyes for a second or two and I wondered if he was asking himself, 'Did I make any impression on this youth last night when I gave him a good talking to?'

In the queue for the cage I tried to hide my apprehension by forcing conversation with Les: 'What's thy lamp number Les?' 'What dist do last night after thaiy gotst thy pit—dirt off?' 'What hast got for thy snappin'?'—anything at all which came into my head which would keep my mind off the cold, dark shaft.

Les ignored my questions for his mind was on other things, he seemed to sense the condition of my mind. As the shuddering rope decreased its speed Les warned me, 'Nagh, dunna forget th' gate Harold. Just expect a loud bang when he begins to wind and then thou wutna bey frittened.'

'Owraight Les. Thanks. I'll watch out for it this morning. It did put the fear of God up me yesterday. If only there were two ropes instead of one, it wouldna be so bad; there'd be another rope if one broke.'

'Eh, Harold, shut up wut. Thaiyt have me getting the wind up.'

BANG, went the gate above our heads as we were hurtled at speed down the shaft.

Before we reached the supporting legs at pit—bottom, there was that awful smell again. I was not in the habit of swearing, but perhaps it came as some means of fortifying my morale, for as we stepped from the cage I said with some strong accent in my voice, 'Les, I shall never, never get used to that bloody stink.'

'Oh, thou wut, thou wut, a few weeks will do it.'

From the long queue of night—shift men waiting to go up came a voice aimed at me with friendly gesture, "ow dost like swinging at the end of th' rope Harold?'

It seemed that someone was aware of my feeling toward that pit—rope. I looked across to the direction of the voice; it was another of my pals from schooldays, Arthur Jones. I always admired him at

51

school, he was carefree and happy. His question and cheerful manner halted my steps for a moment and I shouted with a smile which was not quite genuine, 'It's owraight on the end of th' rope so long as it dunna break, that's all I'm concerned with.'

Les joined in this jovial banter, but as we continued to walk through the pit–bottom shunt, he changed into a more serious frame of mind. 'Eh, 'arold, for God's sake stop talking about the damn pit–rope breaking, we dunna talk about things like that thou knowst.'

I fell into silence for I knew deep down in my mind that the possibility of the rope breaking during the winding of men *was* a fear of mine. How far others entertained this fear I did not know, but I had recollections of having heard during early boyhood that such an accident had occurred and men had gone to the bottom and to their death.

I turned and looked up at Keele engine and as I admired it I felt some pride for the thought went through my mind, 'That's *our* engine.'

Les pointed to his secret hiding place for my jacket. 'Got thy eyesight Harold?'

'Yes, I think so,'

'Is thy lamp–shade on straight?'

'I think so, here, have a look.'

Down we went into the long, dark dip. Les was in charge. 'I'll shout when there's a low girder, Harold. Dunna bang thy head on one of those sods or thaiyt bey as sick as a dog.'

I followed every movement of Les's body, ducking, swaying and then with relief, standing upright when the height permitted it. Suddenly my thoughts went to a man I knew who was a clerk at the 'Co–op' and I realised that the exercise I was having on this journey 'down to my work' was much more than he would have during the course of his whole day of activity.

Immediately we entered the shunt, off came our shirts which were used to wipe away the sweat pouring from our bodies, and the day of perpetual motion had not yet started. Les made sure that I saw how he put his snapping in a safe place. Soon we heard the commanding voice of Arthur Ricket and the signal on the bell—one . . . two . . . three. The first journey was about to start on its way up the dip for the rope was at our end today. I had to make my limbs do their work and silently I questioned Les's prediction that my stiffness would disappear before snapping–time. The day went by, exactly like yesterday. Push, push, push. Hold back, throw in the locker, non–stop. Sweat ran down into my eyes until I felt the salt in my sweat would blind me.

At snapping–time we both threw ourselves down onto the slab of wood without a word to each other. After five minutes we ate our food and as we prepared to restart work I asked Les, 'I'd like to know what is the difference between what you and I are doing in this shunt and what the ponies do in Littlemine district. If this is ordinary work in the pit, what the hell was slavery like?'

My remark amused Les who smiled as he looked straight at me; 'Christ, Harold, thaiyt still as serious as thaiy wast at skew. So thou knowst about th' pones in Littlemine dost?' I nodded and waited for my question to be answered. 'There inna many left in this pit, Harold, and I feel sorry for the poor sods when I've seen them turned out in th' park for a rest in the summer during Wakes Week. They are nearly all blind to a degree for the want of a bit of daylight.'

'And that, Les, is about the only difference between the ponies and us; at least as far as work is concerned; we *do* get back into daylight each day.'

'Oh, 'arold, owd on a bit, there are some miners who, in winter time see little or no daylight. Take thy own faither for instance. He leaves home before thou dost, six in th' morning and if anything goes wrong in Keele district he inna up th' pit again until it has gone dark.'

'Yes, Les, you are quite right there. I have heard my father talk of the old days at Number Six when hours were longer; he has told us that in the winter sometimes he would work a whole week and never see daylight.'

'Ah well, let's hope that we can get out of th' pit one day. But I will say this, this last war *has* improved conditions a bit for miners, at least that is what I keep hearing at home.'

By the end of my second shift I was becoming quite used to lockering the loads as they charged down into the shunt and Les was correct in what he promised, for my stiffness had disappeared. But my hands ached and were showing the effect of two days of heavy, dirty work. Walking slowly up the dip that day Les asked why I was so silent. I explained that I was wondering which would be the best night school classes to attend so that I could learn something about the theory and technique of pit–work.

On my fourth day down the pit, Les Bebbington did not turn up and I got on with the work by myself. No–one offered any explanation for Les's absence. The waggoners realised that I was having a tough time for they brought their waggons just a little further into the shunt. But I was having to deal with *all* the loads and I soon realised that I was doing the work of two boys. I did not complain too loudly for I was a new boy, but on the fifth day I was told that Les had been sent to another part of the pit. I was working like a Trojan, never one minute to stop and at times I let out a loud groan out of sheer frustration when I encountered a particularly heavy waggon or load. Just before snapping–time my father appeared in the shunt carrying his shot–firing battery and long reel of flex. He came up to me just as I had dealt with a heavy waggon; sweat poured from my face and I was feeling sick with exhaustion. Panting and with trembling voice I said, 'Why am I here in this shunt alone? There has always been two boys here. This is too much for one; why has Les been taken away?'

My father looked straight into my face by raising his lamp, but with his usual coolness and command replied, 'Are you trying to tell *me* what labour is required here?' I could only remain silent for I felt very small and ashamed as I suddenly realised the extent of my impertinence. It was more than I dare do to argue. I felt like a private soldier who had dared to tell his General what tactics to employ in battle and I wished at that moment that the earth would swallow me up. Again my father raised his lamp to look into my face; I suspected that he was looking for any sign of tears. Then with a stronger tone of voice which I had heard throughout childhood, he said slowly, 'Yes, there *has* always been two lads in this shunt but there's never been a Brown working in this shunt before thee, and thou art a Brown artna? Inna a Brown worth any two other lads then? I towd thee that much t' other night in th' parlour. In any case, if thou workst by thyself there wunna be any fooling around, no sky–larking will there?'

I was lost for words for I had never had to deal with such a situation before. I regretted having uttered one word of complaint and I turned to resume my work. 'What *have* I done?' I asked myself silently, 'A Brown complaining about hard work.'

My father put down his shot–firing equipment and began pushing up the waggons with me; one, then another, until that side of the shunt was clear. Then he turned to the loads and helped me get the situation back to normal. He picked up his things again, came to my side and, with a distant strain of compassion in his voice, said, 'Look thee here Siree, no–one in this pit or in th' Owd Number Six can ever say that I favoured my own lads and I dunna intend favouring thee. So, just get on with thy work and dunna let me hear any more o' this blartin'.'

I just did not know what to make of it. Here I was, faced with the work of two boys and there was nothing else for me to do but to get on with it, only because my father believed that the Browns were capable of more work than other people. So, I resigned myself to it from the second day that I worked alone in that shunt; just as only four days before I had resigned myself to working in the pit for the rest of my active life. The only advantage of this job in the shunt was that it was 'all days'.

A few days after this confrontation with my father, I joined Redmond D'Arcy and the two waggoners for our snapping–time. Soon we were joined by Arthur Ricket, the Road–doggy. After a short silence he addressed me, 'It looks as though thy father dunna intend Les coming back into th' shunt with thee, Harold. Has he said anything to thee about it?'

'Yes he has and I'm to stick at it and do th' job myself.'

'Ah, thy father's a hard mon thou knowst. He's good at his job as everybody will tell thee, but, by God, he does expect everybody else to work at th' same pace as himself. Why, up at Number Six your two brothers used to complain for they were always given the hardest jobs. It made no difference, thy father would not favour them. One thing I'll say though, he dunna expect anyone to do what he conna do himself.'

These remarks made me feel proud; this man had complained about my father's strict methods, but in the same breath had praised him for his efficiency as a miner and an official. As I walked slowly back into the shunt my mind went back to the day when he had received a letter from the Management of the Minnie Pit which had blown up, killing one hundred and fifty–five men and boys. The letter expressed the gratitude of all concerned for the work of the rescue brigades which had rushed to the scene and had worked day and night until all bodies had been recovered.

Again I felt ashamed that I had complained to my father about my working alone in that shunt. I set my teeth and vowed to myself that I should not utter another word of complaint and that I should do my utmost to maintain the tradition which my father had brought to my notice on the evening of my first day of work in the pit. The second half of the shift began with its roar of waggon–traffic, but I had just heard words which made all the difference to my attitude to my work. The injection of admiration for my father eased the burden of my work a little and helped me to face the bleak future of work underground. I went at my work with a will and the determination to become a useful pitman myself.

Soon I had enrolled on a fresh course of study at night school, a course appropriate to my new work underground. In addition to arithmetic and English at my old day school I had the advantage of another course of lessons at the National Church School at the top end of the village. This was run by our own Pit General Manager. At both these evening schools I saw men of mature age who, after a hard day's work in the pit struggled with their problems and were making great efforts to improve themselves and become better qualified for their work in the pits. To me it was an inspiration to find myself sitting beside them in class. When the winter session ended our own headmaster, Mr Frank Ellams, agreed to teach at a summer night school for one night a week. The subjects at the course run by our Pit Manager were well above my head, but I was determined to forge ahead. My father's books on Practical Mining were at my elbow at home and with his long years of experience I had every advantage. These books became my constant companions once I had eaten my meal, washed off my pit–dirt and performed my necessary practice at the piano. The Davy safety–lamp, detection of gas, brake horsepower of engines, faults in ropes and cables, pulleys, stress, hydraulics, use of timber for support at the coalface, the effects of fire–damp, the cause of gob–fire and explosion, spontaneous internal combustion, elementary geology, etc. It seemed that there was no end to the list of subjects to be studied, even to take the first qualifying certificate, that for the work as a shot–lighter. I became deeply involved in study and with this I was happy, I was learning. Walking up Back Lane to the pit in the early hours of morning I found myself repeating almost aloud the names of the various strata connected with my study of Geology, 'Pleistocene; Tertiary; Permian; Cretaceous; Triassic; Carboniferous; Devonian; Cambrian; etc.'

'If only I could study during the day when my mind and body are fresh and at ease, instead of having to do all this after a day in the pit.' Such thoughts haunted me almost every day and how I envied those boys who, because their parents could afford it, went on to Grammar School, matriculation and then, perhaps, to university. But I struggled on with my studies. There were few facilities at home for absolute quietness for we all had to share what space there was and if I did find a quiet corner it was invariably cold and uncomfortable during winter months.

After about two months of my working in the shunt, my father returned from the pit one day and even before taking off his coat called out to me, 'Here, Siree, I want a word with thaiy. Come in here and I will talk to thee while I eat my dinner. The Gaffer's been having a word with me today about thee and thy night school. He says that you must be given more practical work which is relevant to your mining study. He says that you must be put with a ripper and then a stint or two with a road–doggy and learn all regulations regarding rails, sleeper widths and rules for manholes on main haulage roads. Then thou art to go th' face for a bit to get some idea what colliers are expected to do.'

My head became hot as I realised that there were still other subjects to add to my long list in the study of mining. I said nothing, but waited for my father to continue.

'Then, thou't have to go and work with a bricky and learn how to lay bricks and build those heavy doors to secure ventilation to the far ends of districts. That is most important thou knowst. You must realise that all this practical experience is necessary and before thou canst sit for thy Fireman's papers thou wut have to work on th' face for two years.'

'Yes, but work on the face will not be for a long time yet will it? Working on the face means loading and that's a man's job isn't it?'

'Ah, it *is* a man's job on th' face, but thaiyt a Brown and thou canst do it. Artna thee a mon then?'

'Does it mean that I shall have to leave the shunt then?'

'In about two weeks time. I shall then be putting thee with a ripper, Harry Cleer. He's working on a rip in the Keele air–road. It will be heavy work, but thaiyt do it owraight.'

I had already learned that ripping was hard work, for rippers worked on contract and were paid for the distance they ripped each week. Yardage was a very significant word in their pit–language.

Also, they ripped down *dirt* and not coal; dirt was nearly twice the weight of coal and it had to be loaded with a shovel. I had heard others complain of the back–breaking misery of loading dirt. Only men who were not afraid of work accepted the job as 'ripper'. I had no doubts as to what kind of life I was in for with a conscientious ripper.

It was during these last two weeks in the shunt before going to work with the ripper that I had a nasty experience. It was a revolting scene which has never faded from my memory; it made a deep mark on my subconscious mind. There were very few minutes during the day–shift when coal drawing stopped, except for snapping–time. If a journey came off the road or if there was a fall of roof, or drawing had to stop for an accident, then we all sat around and indulged in idle talk. One day, just after snapping–time the main–dip journey came off the road and we all waited, talked and joked among ourselves. We were joined by Arnold Wardle, who, becoming aware of the stillness, walked up from his pump for the sake of a few minutes conversation. Arnold had lost a leg in an accident at Number Six Pit and had been given this light job as part compensation. It was Arnold's light which I had noticed during those very first minutes when I had arrived at the bottom of Keele dip on my very first day of work in the pit.

We all enjoyed our harmless fun together until we were joined by another man. This man worked near the far end and had come out to investigate the long wait for waggons. From the never–ending clangour and bustle of coal–drawing to this deadly stillness created an eerie, unnatural atmosphere. There was something pleasant about it to my young mind, until the arrival of this newcomer. He was known for his vulgar language and his uncouth manner underground. When men work together under harsh and dirty conditions, swear–words do come easy on the tongue and it is excused; it is a spontaneous explosion of words and the violent moment is soon forgotten. But this new member to our cheerful company of pitmen seemed unable to say even a few words without the use of a very common adjective which not only colours the swear–word vocabulary, but it is a word which also degrades the function of speech. Such words belong to the very extreme in bad language.

'What's the bloody matter then? Has the—pit closed for today?' He looked around in a furtive manner as if searching for some devil–may–care mischief; he began throwing bits of dirt at the waggoners.

'Eh, cut that out wut,' shouted back one young man.

The newcomer swore, but continued with his tomfoolery and soon I received a sharp sting on my face from one of his bits of dirt. Hearing the sound of my voice as I objected to the possible danger of a piece of dirt striking me in the eye, this uncouth man laughed and shouted out loudly, 'Bugger me if it inna Harold Brown. So thy faither's got thee into th' pit then. Following in the steps of thy two brothers art then? Another Brown, another pitmon in th' Brown family. Eh lads, 'as 'arold been entered into th' club yet?'

I had no idea what all this meant, but I was filled with apprehension. There was no movement until this man from the far end rose and came towards me with the remark, 'Eh, come on lads, we conna 'ave a new lad in th' pit without him being entered into th' club.' Only one other came to his assistance as I was held fast. I lashed out with my feet and arms and let out hysterical screams as they struggled to loosen my leather strap. It was a degrading and humiliating scene and the details of this vulgar initiation remained with me throughout life. It took me some time to regain my composure, but when I had gathered my wits about me again I wandered back to the bottom of the main dip. No–one seemed to notice me until the man who had instigated the vulgarity shouted to me, 'Well how dost feel now that thou art a member of th' club?'

That was too much. I quietly hung up my lamp and I made sure that his lamp was safely out of the way, then I threw myself at him full–length and there was a scuffle, but Redmond and others soon had us apart. I was too young to know that fighting in the pit was strictly against the rules.

'Bacca on th' bonk' I shouted at my assailant. He had started this ugly business and now, in the language of pitmen, I invited him to meet me on the surface at the end of the shift.

I knew I should lose for he was an older and stronger man. I was not without experience for my eldest brother Edwin had tought me how to take care of myself. Redmond D'Arcy came to me and told me that he did not think my opponent would turn up behind the boiler–house on the surface. 'I have warned him of the consequences if your father hears of it. Also I have reminded him that he would have your brother Edwin to deal with afterwards and you know what a fighter he is.'

'Well,' I said to Redmond, 'I shall go just in case he does turn up. Would you come and second me?'

'Yes I will, but I think you will find that he will not be there.'

Redmond was right for we waited a while, but no one appeared and that satisfied me for I did not relish the thought of the fight. Redmond remained silent for a while, then said slowly, 'It is unfortunate about that long wait for the journey for this would not have happened if there had not been that fall of roof near the pit–bottom. It shows what a world of chance we do live in Harold, but try to get it out of your mind and learn from the experience.' I was full of gratitude toward Redmond who had come out on my side. How I admired this young Irishman who held his principles before him. No rough language, no rudeness or lewd jokes, no bawdy stories about sex or women.

I turned away from the pit that day and walked slowly through the cleft which divided the two high dirt–tips and, as I reached the railway bridge, there standing waiting for me was my Sunday School friend, Bernard Cliff. There had been three of us at Sunday School, always together and always very close boyhood friends; the third boy was Eric Fairbanks, a cheerful lad, and liked by everyone. Poor Eric had started work at Knutton Forge as an oil–boy and, like Bernard and me, he had left school at the age of thirteen. One day he was oiling the bearing of the huge flywheel while it was still in motion. The end of the shaft became engaged in the front of his overall; it twisted the material round and round, throwing the lad into the middle of the whirling monster. What a shock it was, all his schoolmates took it very badly for we all still young boys trying to get our feet planted on the first rung of the ladder in life. Eric's death was a dreadful tragedy and it seemed to throw Bernard and me even more closely together.

As Bernard approached me on the bridge I tried to produce a smile, but my countenance no doubt reflected the condition of my mind. 'What's the matter Harold? You look down in the mouth,' said Bernard showing some concern. I did not reply at once, but we both turned into the lane which led up to the cricket field, sat down on the grass verge and I related to him the unpleasant incident of the day.

Bernard shared my disgust, 'Harold, isn't it awful? Why, they do the same kind of thing up at the brickworks. Most lads are expected to be entered into some kind of club and I suppose I was just lucky for they never got a hold of me. It is cruel really because some kids are rather nesh and timid.'

Bernard, like Eric and myself, had taken whatever job came along in order to earn money and help the family purse. But Bernard had escaped the brickyard and now was established in better work in a dairy shop in Newcastle–under–Lyme. A close friendship was being established between us and at this moment Bernard realised my unsteady state of mind and suggested that we should meet that evening and attend the first house at our local cinema.

I did not mention the scuffle, nor any of the events of the day in the pit to my father. He did not discuss matters concerning the pit during the hours that he was at home, unless there was something directly concerning me. I was expecting him to ask me questions about the long period of 'waiting for waggons' when the fall of roof had occurred. Deliberately I waited at the table until he had finished his last morsel of food and taken up his daily paper. I wanted to be quite sure that he had not been told of my coming to grips with another man underground.

My last day of working in the shunt arrived. Already I had learned a lot since my first day with Les Bebbington and as the hour for 'loosit' approached I gazed at that broad flat roof, that solid slab of rock which gave such a sensation of 'safety'. It was very low, but it was solid and I knew that it was safe; I became aware that it had been a source of comfort to my young mind. No one knew just how far to the left and to the right this great slab extended; only I knew just what assurance and peculiar sense of pleasure I had derived from it when I compared it with the uneven, jagged and unsafe appearance of the roof in other parts of Keele dip. That even roof of the dip–bottom had proved to be the very antithesis of the danger I felt when hanging at the end of the pit–rope with the deep dark shaft below me. I knew that I was going to miss the mental assurance which this solid, unbroken roof gave to me. Bottom of Keele dip was a safe cavern.

4 Ripping dirt and waggoning

'Wait at th' top of Keele dip for Mr Cleer. He will take you to his rip.'
These were my father's only instructions as he tested my lamp on the
Monday morning on which I had to begin my new work. All the
weekend my mind had been at work asking, 'What will my next job
be like. What will Harry Cleer be like and am I strong enough to load
dirt? Will there be a safe roof?' Of one thing I could be certain; Harry
Cleer would be a hard worker, for my father would not be putting his
son to work with anyone but a good workman who knew his job well.

I watched the day–shift men stand for a few minutes to get their
eyesight before entering their own district. At length, a short,
thick–set man approached who looked at me from the corner of his
eye; it was almost a squint and at once I knew that this was my man.
As he took stock of me from the corner of his eye I began to wonder,
'Has he got the stag [nystagmus]? Has he worked on the face for years
and become a victim of the dread eye complaint brought on by
constant eye–strain under the glimmer of the Davy lamp?

'I felt sorry for him already because of this squint. He carried a
pick–shaft on which were three freshly–sharpened blades. Again he
looked at me with that sideways glance and said sharply, 'Art thee
George's lad then?' I nodded and replied, 'Yes, I suppose you are Mr
Cleer?'

There came no reply but after a few words with another man he said
with a stern, almost unfriendly tone, 'Here lad, get owd 'o these picks
wut; we are workin' in th' air–road, through that there door.'

I had been told of the function of the air–road by my father and I
was now anxious to know what it was like. The air on the other side of
this door was not far from the upcast shaft; there was terrific 'pull' on
the other side and I found it very difficult even to move the door at all.
Cleer saw my difficulty and immediately he threw down his coat and
said, a little more kindly, ''ere lad, let maiy show thee how to do it.'

A minute ago he had barked out a sharp, almost unfriendly
command, but now, it was a gentle offer of help to an inexperienced
lad. I did not like his first order, but it was more than balanced by his

readiness to help me when he saw my hands struggling with the heavy door. He put his foot on to the side of the door frame and pulled with all his might. It was not easy even for him; he turned to me and, still without a sign of warmth in his countenance, said, 'Got it lad? Canst see th' knack? Thaiyt get used to it, but thou't have to learn to open th' doors thou knowst.' All this accentuated my inexperience and I felt like an apprentice learning a new trade.

We passed through two of these heavy air–doors. At once I asked, 'Mr Cleer, why are there two doors here?'

'I shanna have time to answer thy questions all day thou knowst, but since thou hast shown some interest I'll tell thee. Double doors prevent any short–circuiting of air from the main inbye road back to the upcast shaft. Double doors maintain ventilation to the far end workings. Is that enough for thee?'

'Thank you Mr Cleer; that's very interesting.'

We could now not only feel the air, we could hear it rushing toward the upcast shaft. Cleer also explained that the air–roads are escape routes for the men when a fall of roof blocks the main road. The roof was very low in places and at some points we had to crawl along on our bellies, pulling our bodies forward an inch at a time. It was a frightening experience, for the air was compressed into such a small space, giving the impression of a whirlwind. Once we were able to stand upright again, I observed the heavy panting of Cleer. Taking a moment to get his breath he asked, 'Eh, what's thy name then? I know thy brother Ted and I did know thy other brother who was killed in th' war. George wanna it?' I nodded and before I could reply he went on. 'And my name is Harry Cleer. Just call me Harry.'

He was silent for a few minutes. The crawling had taken toll. 'I hope thou't a good worker, well thou must bey, thaiyt a Brown and there anna any idle Browns on Silverdale.' I felt cheered by the compliment to my family and replied, 'My name's Harold.'

We continued our journey and soon encountered an old man with a beard, a man who did odd jobs of repair work in the air–road; it was old Jack Mathas. Already he had his coat off and held a tool in his hand. His mate stood nearby; it was Paddy D'Arcy, brother of my friend Redmond. He laughed as I came near to him and I lifted my lamp to his face. Paddy was a cheerful young man and, like Redmond, as strong as an ox. All three D'Arcy brothers were of cheerful nature, they worked hard and were well–liked. To me their musical Irish

63

accent provided a pleasant relief to our own familiar Staffordshire manner of speaking. Paddy made a few helpful remarks flavoured with a note of warning regarding the loading of dirt. I felt less apprehensive with the knowledge that this happy young man would be working only a few hundred yards above our rip. Another length of road lay before us about two feet square through which we had to drag ourselves and there before us was the rip. As I raised myself to an upright position my eyes fell upon the empty waggon, a large shovel and 'the dirt'. Cleer hung up his lamp and nodded in the direction of the dirt and told me to get started.

I took off my shirt at once, but Cleer came to me with the warning, 'Nagh, young Brown, look thee, dunna take off thy shirt immediately. It is cooler here in this air–road compared with bottom of Keele dip and further inbye. Wait till thou art warmed up a bit, then take thy shirt off. But, when thou hast to stop for more than a few minutes, be sure to put it back on again, dost understand maiy?'

'Thank you Mr Cleer,' I replied, feeling pleased with his concern for me.

' 'ere lad, my name's 'arry, hast forgotten?'

With this more friendly atmosphere established, I felt a little better disposed toward the great pile of dirt before me. However, my pleasure was short–lived for within minutes the great heavy shovel and the dirt began to pull at the muscles of my back and arms; they were informing me that this was a different class of pit–work to that in the shunt. But I was not dull of heart for Harry Cleer had shown that he was concerned for my welfare; he need not have mentioned the cooler air in the air–road. He was a cheerless man, almost completely without a smile. He had, up to the present moment, exhibited no sense of humour at all and yet he had warned me to take care, he had tried to make sure that I should not come to any harm. He stopped his picking at the hard rockface above his head where he prepared for a cross–bar to be put into position. He had seen me walk over to my shirt which I used from time to time to wipe the sweat from my face. I turned to assure him that I knew what I was doing. 'It's owraight Mr Cleer, but I'm getting hot, the sweat keeps running down into my eyes. This dirt is damned heavy, how many loads a shift do you have of this?'

'Never thou mind *how many* loads a shift, just keep loading it. Just keep on chucking it in, thou't get used to it by snappin' time. Keep thy shovel swinging like a machine, try to think that thou art part of an engine, that'll help thee a bit.'

I could have forgiven him that remark and the attitude which went with it, had it been meant as a joke, but Harry Cleer was quite serious about what he had advised me to do. And so it went on, shovel, shovel, shovel. My back was breaking and my shoulders and arms ready to drop from my body. The utter slavery of it all was relieved only in a slight degree when from time to time I straightened my back, looked across at Cleer and let the sound of his voice re–echo through my mind, imagining that he really wanted to share a joke; but I suspected that he considered such a facetious state of mind too much of a contradiction to the seriousness of his ripping. 'Keep thy shovel swinging, try to think that thou art part of a machine . . .' Indeed!

At long last it was snapping–time. No rings on the bell here, no loud voice of Redmond shouting 'Snapping'. Harry looked at his watch and announced quietly, 'Snapping–time, lad; get thy jacket and shirt on, now we've got to get along to Plate Landing.'

'Plate Landing? Plate Landing?' I asked, showing some concern, for I was famished. 'Anna we going to have our snapping then?'

'No, not until we have put our loads of dirt on to th' journey in th' main dip.'

My heart sank for hunger was getting the better of me and I was beginning to rely upon my snapping to restore my strength. I followed him up the incline, then turned right where our loads of dirt stood. Then, just as we had done in the pit-bottom to enter the air-road, we passed through two heavy air-doors and into the main dip. Plate Landing I knew well from my travel up and down the dip. Now I could see the significance of those two heavy iron plates which stood, hinged on to both sides of the track. The rope was swinging up and down and I could see from the strain on it that the journey of loads was on its way up. This gave us a few minutes to *wait* and, oh, what luxury that was. Now came the rumbling of the journey. Plate Landing was right on the most steep part of the dip and I marvelled that the rope could take such a strain and heavy load.

Cleer again looked at his watch. 'Just about right Harold, it takes only a few minutes for him to land, turn in again and come back to us. Nagh, just watch maiy, what I do when he comes back, but dunna ask questions for we have only the twenty minutes of snapping-time to get our waggons off th' rope and our loads on to th' journey.' I said nothing, but wondered why I could not have been eating a bit of my snapping during this short waiting period. I helped Cleer to put the safety beam across the track and lower the heavy plates; this formed a flat platform. Cleer now became alert and took his lamp in his hand, holding it near the bell-wires. Here it came, our empty waggons at the end of the rope, approaching only at an inch at a time. The loading was all over in ten minutes and I marvelled at the skill of my ripper-mate. I even forgot my hunger for a short while. Cleer had given distinct signals to the engineman by using the top half of his lamp to make contact on the bell wires. Each movement had been slow and gentle. The expert handling of this operation on this very steep part of the incline; the complete understanding of Cleer and engineman by way of this crude means of signalling; it all fascinated me and when I told him how interesting it all was, he just turned and said, 'It's all in a day's work, 'arold lad. Nowt to it really, but thou now knowst why it is called Plate Landing.'

For our snapping-time Cleer chose a spot just behind the air-doors, not in the air-road itself. 'It's a bit warmer here lad.'

I watched him put his coat around his shoulders, so I assumed that it was the wise thing to do and followed his example. Immediately I took a long drink from my bottle.

'Eh, eh, steady on thy water lad. Thaiyt regret it. I towd thee about that early on when thou tookst thy first swig at thy bottle. Save some for the end of the shift when thou has finished thy sweating.'

I opened my snapping carefully and looked for vermin.

'Thou dostna have to worry about cockroaches and mice in th' air-road Harold. It's too cold for them in this air.'

'Who was loading your dirt up to last week Harry?' I asked, at last daring to call him by his first name.

'Oh a chap from Newcastle. He was going to get married and th' money on this rip wanna enough for him. He asked thy faither for a job loading on th' face to bring his money up a bit.'

'He was a man then, an adult?'

Cleer looked straight at me; he had let the cat out of the bag. He hesitated. 'Ah, he *was* a mon. A big chap for his age, though, and he loaded dirt very well, but I conna pay a loader. I could pay him only th' basic rate and th' Gaffer knows that I am obliged to have a mate working in this air-road. So, thy father thought it a good idea to let thee see a bit of the ripping side of the work in th' pit.'

'I see, Mr Cleer, so really I am doing a man's job with you?'

My question remained unanswered for Cleer no doubt realised that I was wondering why I, a lad of fourteen, should have taken the place of an adult on this job of heavy work.

Obviously keen to change the subject, Cleer remarked, 'I wish thou wouldst call me 'arry. Try to remember, lad. I dunna like this Mr Cleer all the time, but I do like thy manners.'

I fell into silence; I could not help reflecting upon the fact that I had worked alone in the shunt where two boys had usually worked, I was now loading dirt, replacing a grown young man and I began to wonder if there would be an end to the handicap of my having been born a Brown.

Immediately the prescribed twenty minutes had elapsed Cleer again took out his watch, shook himself and snapped out an order in a manner calculated to remind me that I was only an apprentice, learning the trade, "ere, go up and get th' drilling machine. It's up at th' top of th' rip under some timber, thaiy costna miss it.'

I brought down the heavy machine, its weight pulling on my shoulders.

'Hast done any drilling yet Harold?'

'No, I have never had a machine in my hands before.'

Cleer then showed me how to erect a post to support the machine. 'Nagh, lad, this post inna to support th' roof. It is to take the weight of the machine. Dost think thou canst do it? Keep thy hand on th' post and try not to let the machine slip or thou't have the weight of the whole thing on thy foot and *that'll* make thee yelp. This is called a ratchet-drill, dost understand maiy?' Patiently Cleer showed me how to use this heavy drilling-machine and explained the difference between this equipment and the light machine which I might one day encounter on the coalface. Like all beginners I made mistakes, but once I felt the drill biting into the hard rock I knew that I had the knack. As the drill gradually disappeared I felt an inner satisfaction. My first hole drilled in my pit-life, and I had drilled it into hard stone. I experienced a feeling of gratitude to my teacher Harry Cleer. Hardly had I taken down the machine than I saw another light appear above us.

This was my father arriving to fire the shot. He had used the snapping–time to walk up the lower part of the main dip below Plate

Landing, taking advantage of there being no traffic during that twenty minutes. He merely nodded to me as he walked across to Cleer. 'Well Harry ,what's he like, can he use a shovel?'

'Ah, he's owraight, George, let him alone, he'll learn.'

My father's visit was welcome; I had spent a whole morning in the cheerless company of Cleer and I knew that, at home at least, my father had a sense of humour. Also my father's presence made it impossible for me to load the dirt, so that meant a short rest for me.

I stood there watching and I was fascinated at the care with which my father placed every piece of ramming into the shot–hole, then gently rammed it right home on to the shot–powder with a short pole. Then the connecting wire and flex from the explosive charge were carried to a spot as far away as the flex would allow. I watched every move in a state of awe; this was my first experience of shot–firing.

'Look, lad,' said my father as he held his firing–battery firmly in his hand and well away from the loose ends of wire at the end of the flex, 'once thy powder is rammed and the detonator in position, thou must always keep thy mind on what thou art doing. Shot–firing is a serious business.

He then remained silent as though he was giving my intelligence a chance to work out what came next.

The nature of the instruction also seemed to suggest that one day I should be doing the same kind of work my father was now doing. Again he called me. 'Nagh, look thee Harold, I've sent Harry Cleer *down* below the rip to stop anyone coming *up* the air–road, I shall go *up* the road to fire and by doing that shall be able to stop anyone approaching the shot from that direction. Now, where has *thou* got to go?'

I scratched my head and pretended to be thinking hard and he tried to prompt a reply from me.

'Well, come Siree, there's only one other road into th' place, inna there?'

'Oh yes, along to the air–doors leading to th' main dip', I replied at last.

'Ah, that's right. Now, go out to the main dip and dunna allow anyone, not even th' Gaffer, to come along that road till I've fired. Just say, "Firing", and if it is th' Gaffer, remember thy manners. This is a strict law and very necessary precaution when firing. All roads must be guarded, see that thou keepst that in mind.'

'Only seconds after I had emerged from the air–road on to the main dip I heard the dull 'thud'. I knew then that the shot had been fired.

The door seemed to shudder a little from the impact of the explosion. Cleer came to tell me that all was ready for my return. My father was carefully testing the roof with the end of a pick–shaft. I stood waiting and admiring him as he tapped the roof gently and listened. There was no hurry, no chances had to be taken with any of the operations connected with shot–firing. The place was still thick with dust and I felt my throat dry up as I took it down into my lungs. I observed how both my father and the ripper coughed and spat up the dust from their throats. This was the deepest impression which firing of this shot, the first in my experience, had made upon me; the dust, for I could see for myself now why so many miners suffered from the dreaded dust disease. There at my feet was yet another great pile of dirt to be loaded. The sight sickened me, but there was nothing for it but to reduce the size of the pile, a shovelful at a time.

It went on and on, shift after shift, week after week, one day the same as any other, no variety, monotonous hard work, and I would have given anything to have seen my proficient ripper–mate, Harry Cleer, break out into laughter or song. But I did realise that I could not expect to have everything. He was teaching me things and I admired his skill. I soon mastered drilling with the ratchet–drill and I did try to pretend that I was part of a machine in my attitude to the dirt and the shovel, but the dirt never became lighter in weight. I was a young man who enjoyed a good joke, a hearty laugh from time to time, and I was accustomed to singing or whistling quite spontaneously. But here, all alone with Harry Cleer, all that exuberance would have been out of place and I felt that my whole being was being cramped, life was dull and soulless. But I respected this expert ripper, this good pitman and his industry. I reminded myself that not everyone in the world was born to be light–hearted and happy. Cleer had a living to earn and his own task was to rip the air–road and make the roof safe; this may have given him all the satisfaction he required from life. I often wondered if he had domestic troubles or perhaps a very sick relative. How was I to know that perhaps in his heart he was singing, 'How far can we rip today? More yardage means more pay.'

Each Friday my father arrived to measure up so that Cleer's wages could be calculated for the payday in a week's time. What Cleer earned meant nothing to me for I was paid at the flat rate. I gathered that the ripper was not doing at all badly, but it was against all the rules, (especially in our house) for me to ask if I ought to receive a little in the way of encouragement. That would have been a breach of good manners, and we had been well–drilled in that subject since the day we could understand such things. When I felt like complaining, my father's words rang in my ears, 'no blarting'. The dull monotony of this job in the air–road, the very intensity of the boredom and my having no other human being to speak to but Cleer, made relief all the more welcome when at last it did come. A cheerful ripper with the nature of a comedian may have relieved the boredom, but he may not have taught me anything. Harry Cleer had done that, for I left the rip with several minor skills added to my ability and I was grateful to him for that. My memory of the hard work, the weight of dirt on the shovel, the misery of long weary shifts alone with a cheerless man were all tempered by the awareness that Harry Cleer had done his best to show me 'how it was done.' No doubt Cleer had fulfilled his contract with my father to show me a bit about the ripping side of pit–work.

I did not notice my father enter our living–room for I had my head resting between my hands; my mind was grappling with a mathematical problem contained in my homework from night school. Suddenly his voice broke the silence. 'Thou't be waggoning on nights from next Sunday. You will be on the length between four's and Seven's dips.'

'And where is it then?' I asked, with some apprehension.

'You know Number One dip at the top of the shunt?' I nodded silently, for that was as far as I had explored above the shunt. 'Well, it is further on up that road going toward the far end.'

My mind left my study and I wondered, 'If I were not a Brown and if my own father was not my boss at work, I should choose whether or not I changed my place of work. At least I should have some say in the matter.' I looked across the table at my father and wondered if he ever thought to ask me whether I would like to go waggoning or to work on a rip, or to go up and work on the face. I was about to say what was on my mind, but prudence gave me a nudge and I held my tongue. Also, these deliberations in my mind were serving to bring some relief for this news meant that I should be leaving the rip.

I had worked all days with Harry Cleer, but now came 'nights'. I did not go to bed during the day before my first night as a waggoner. It was Sunday and I had slept all of Saturday night. I walked down Back Lane and into the road which led to our Sunday School for I knew that Bernard Cliff would not miss that if he could help it. I listened to the hearty singing and Bernard and I enjoyed a quiet walk. Quite an ordinary experience, but the pleasure of it remained deep down in my mind. After supper I made my way slowly to the pit and, passing a house in which a group of people were singing hymns to the accompaniment of piano and violin, I stopped and listened for I felt myself envying them their homely pleasure. I walked by the fustian factory and thought of all those girls who worked there and pictured them all at the Bible Classes that afternoon; then the miners, steel and ironworkers who would have been standing on the opposite side of the gallery in the Sunday School. All this occupied my mind as I walked quietly to the pit; and it all gave me a very agreeable feeling.

The pit was usually poorly attended on Sunday night and I was alone as I walked slowly down the dip. I could not dismiss Sunday School, the music coming from the house and the strange feeling within my mind as I dropped my feet into the sleeper–holes. There was an awareness that I was leaving childhood behind me and perhaps entering at too early an age the hard life of a miner with its grim conditions and its darkness. I arrived at Four's dip early, then took the trouble to walk the whole length of my stretch of road up to Seven's dip, merely to try and assess what would be required of me. It was a low roof all the way and as I arrived back at Four's my back was already aching. I sat there waiting and wondering if I should measure up to the hazards of this new job. I recollected that I had seen some of the waggoners and that they were all above me in age.

Soon there arrived two young men of seventeen or eighteen years of age. They stripped to the waist and looked at me. One smiled and came over to me, raised his lamp and asked, 'Ah–do, art thee Harold Brown then?'

'Ah, that's raight. I'm supposed to be waggoning between Four's and Seven's.'

By a nod of his head he beckoned me to follow him. 'Come with maiy, lad, thaiyt owraight, this is it. Come on I'll show thee a bit of th' ropes. Bring thy lamp, thaiyt 'ave to keep thy light at half–way. Thou costna carry thy lamp on this length, too many bloody deep sleeper–holes. We push in, one after t' other and we dunna go out till we are all three up at Seven's dip. Dost understand kid?'

'Yes, I think I've got it,' I replied taking in deep breath. 'But dunna that mean that we are all working in th' dark?'

'Ah, it does lad except at each end of th' length and that bit in th' middle where thy lamp will be hanging. But dunna worry thysel' kid, thou't know every pairs o' rails and every bloody sleeper–'ole before snapping–time. 'ere, thou't a nipper for this waggoning length, didst ask for it then? Thou wutna get any more money thou knowst working on th' turn like this.'

Already I was feeling exhausted by the low roof and the journey with bent back over the length for a second time since I arrived. Down I went on to one knee. 'No, I didna ask for it, my father sent me here.'

'Thy faither, thy faither, oh, I see, thaiyt George Brown's lad then? What job hast been on then?

'In th' air–road with a ripper.'

'Hast now, a nipper loading dirt! God I had a couple of days of that; no more of that for maiy, it was a weary job.'

This young man, to whom I already was feeling grateful for his instruction and advice, and even more for the sense of humour which was being interpreted by the very tone of his voice, was Bill Fry. The contrast to the mood of Harry Cleer had already registered in my mind. This young man was 'real North Staffordshire, real Potteries.' I appreciated those who could laugh at things and make the best of difficulties. The other waggoner I had seen before when I worked in the shunt. He was Johnny Cameron who had walked behind Fry and me as I had been 'shown a bit' of th' ropes'. Cameron said nothing, Bill Fry seemed to be in charge. We were all three now at Four's dip with a lamp at each end of the length and one lamp at the half–way mark.

As the two experienced waggoners took hold of their first waggon of the shift, Cameron took me aside and said softly, 'Follow us in gently at first. We'll wait for thee, but look thee 'ere kid, keep both hands *below* the top of thy waggons and loads. There inna room for thy fingers on top. Waggons and loads have to be held in th' buffers. Thaiyt get used to it in time, but it's bloody hard work. Thaiyt a nipper and thou must try and stretch thy hands over th' buffers to owd thy load steady. Dunna try to look over th' top of thy load on the way out or thou't bash thy head in. The roof is low all the way down to the next length and if thy load races away with thee it will run down to the waggoners below. If thy load does race away, give a bloody good loud 'oller so that they have a chance to get out of road.'

That was a long speech for the one who had remained silent, but I replied with one word, 'Thanks', and I tried to indicate the extent of my gratitude by the tone of my voice. Already both these young men had gone out of their way to give me useful advice and to point out the pitfalls in the waggoning length. They did it before starting their arduous work and I never forgot it. How easily, learning that I was the son of an official, a strict official, they could have let me learn by my mistakes. The warmth which was created within me by their willingless to help a newcomer helped me over the rougher aspects of this first shift.

I bent my back and put my body behind my first waggon, stretching my small hands across the buffers. My fingers hardly reached from one side to the other, but I had to do my best and hope that when I brought down my load, I should have enough grip to hold it back and stop it from racing away. When I arrived at the top of the length with my waggon I was already out of breath. The other two waited and insisted that I go first with my load. No wonder, *they* were not going to risk being in front of a new lad and the possibility of a race–away load bearing down on them.

The generous Bill Fry came to again and said, ''ere kid, when thy load begins to move down that bit of steer in the middle of th' length where thy lamp is hanging, just twist thy load like this.' Bill held the buffers of my load and, using the weight of his body, twisted the load so that the inner flange of the wheels were rubbing against the side of the rails. This acted as a kind of brake and checked the speed to a degree. As my load did gain speed I put Fry's advice into operation and I felt the effect right away as the flange of the wheels rubbed against the rails. How grateful I felt towards Fry and unashamedly I said aloud, 'Thank God for Bill Fry and his kind.'

My worst enemy was the sleeper–holes and I endorsed with my whole heart the brightly–coloured adjective which both these young men had attached to the noun 'sleeper–holes'. They had mentioned the word sleeper–holes only twice, but each time their voices had told me the whole story and put me on my guard. They were quite right, I did know every sleeper–hole by snapping–time. Working in the dark made me aware of the slightest variation in gradient, exactly how many steps there were before a certain deep sleeper–hole and how many steps before the race section. I became increasingly aware that if I let my load go, it would race down and possibly trap some waggoner on the length below. 'What a responsibility for a small boy of fourteen,' I said to myself as I prepared for the second half of this gruelling shift. But I had two good mates; that made all the difference.

I had heard some highly—coloured language in the shunt and at the bottom of Keele dip, but these two young waggoners seemed to possess a vocabulary of swear—words which out—stripped anything I had heard hitherto. There was a good reason for it; these men had to face difficulties and frustrations which would have tried the most saintly of saints. I did not like swearing and had given way to it only a few times since coming into the pit; and even then it was an attempt to boost my manhood and to try and impress those around me and assert that I was one of their company. I still embraced within my mind the efforts of those at our Sunday School who tried to set an example of decency in living and a clean manner of speech before us, as we grew up from infancy. The recent vulgar obscenity which had humiliated me at the bottom of Keele dip and all the filthy language which went with it nauseated me, but the swearing of these two waggoners was another matter. Loads often came off the rails, there was a lot of heavy lifting and mauling because of the roof being almost flush with the top of the waggons. Waggons arrived from the surface not properly greased; loads came down with too much coal above the top edge of the waggon, making it wedge into the roof. The agonising groans, prayers transposed into profanity in an appeal for more strength to deal with the trouble; this kind of swearing I could understand. Non—stop physical effort in an atmosphere which makes any form of clothing a burden; no flow of air, a roof only the height of the waggon and the only light a glimmer from a pit—lamp; lamps placed at long distances apart.

By the end of this first shift I had become used to working in almost complete darkness. My feet knew every sleeper and my ears became accustomed to every joint in the rails as my waggons and loads passed over them. It was a murderous occupation but, oh, how grateful I was for the good nature and character of these, my two mates. Without being fully aware of the significance at that moment, I was witnessing the very essence of the spirit of comradeship of miners at work. One full of fun and vociferous expression; the other, content to follow, almost silent in his mood until the strain brought breaking point and he too joined the chorus of expression; both men hard workers and ready to help at every turn. Two men unlike in many ways, but both displaying the same features as those rowdy waggoners at dip-bottom.

Our work included a lot of shouting on this long length for at intervals we had to wait for the young men taking off the loads at the intermediate jig-dips to bring them down to the bottom of the length; then they would take in their waggon with us. One period of waiting was particularly long. Tim Jones, taker-off at Six's dip, had shouted out his warning that he was 'coming out'. It usually took only seconds for him to appear, but after four minutes or so Bill Fry became impatient and groaned, 'Oh, come on Tim, for God's sake.' He knew that an accumulation of waggons behind us brought trouble, traffic would become halted and we should face a good cursing from the Road-doggy. After another few minutes Cameron became uneasy and said with calm voice, 'I wonder what's up, off th' road perhaps.' ' 'ere 'arold, thaiyt th' youngest, pop up and see wut? said Fry. I accepted the warmhearted tone in Bill's voice and was only too pleased to respond.

Reaching Tim's door I opened it by pulling it towards me. Tim Jones almost fell back into my arms and he shouted, 'Oh owd on, owd on, my leg is trapped.' On the right a post had been knocked out by a lump falling from the roof. The lump and the post had wedged the load and twisted the end of the load which Tim was holding. His right leg was trapped under the buffer. He could not move even an inch without risk. 'My God, Tim, what hast done? Let me try and get thee free.' At the same time I shouted for help with a volume of voice I did not know I possessed.

'Dunna move my load, Harold, dunna move it. If anything moves it will bring down that great lump and all the other stuff.'

My two mates were up in a jiffy and they held the load steady while I withdrew Tim's body from the load. Tim yelled with pain; he was not seriously hurt, but it could easily have been his death. The two waggoners arranged a means of holding the load, and both worked furiously till the load was secure, then they came over to help the injured lad who was holding his leg. Bill Fry placed his arm around Tim's shoulder and said with gentle tone, quite the reverse of his usual vociferousness, 'Nagh Tim lad, lie back on my arm and make thyself comfortable. Dunna worry about a thing, wey'n look after thee.' What a transformation this was, all in a matter of minutes. The contrast to the shouting of abuse to the gods.

Bill Fry laid Tim across his knee as a mother would a babe and I found it difficult to believe. Looking down at the injured lad Fry said, 'Thou't a lucky lad, Tim, that lump was not meant for thy back.'

Cameron returned with the First Aid man and the official in charge. One went straight to the patient; the other turned his safety-lamp up to the offending roof and we all three did the same. There was silence till Bill Fry ejaculated . . . 'Hmmmmmm. Ah, there it is, another big bugger ready to come down.' The official nodded his agreement and immediately gave orders to the colliers who by now had come down from the face to investigate the silence. None of us three knew Tim apart from the fact that he was a 'taker–off '. Immediately my mates knew that he was in trouble, though, their whole attitude changed and their concern was only for the boy's safety. It had to be seen to be believed. I had suffered shock and each time I pushed my waggon past Tim's dip, his trapped leg came before my eyes. Not long after this another man from Silverdale was killed at the Stone Pit and this was followed by yet another death at one of the pits further away.

I had been brought up in an atmosphere of pit–language. Always in my ears there had been reports of burst fingers, blows in the eye at the face, broken backs, cracked skulls and sudden death. The Diglake Flood Disaster of 1895 when seventy–seven men were drowned down the pit; the terrible Minnie Disaster in 1918 when one hundred and fifty–five were killed; the sorrow from that was still over us for it was only three years before. This topic was common at our meal table; my whole growing–up life was conditioned by the harshness of pit–life and the degradation which went with it. Now I had experienced myself Tim's near escape from sudden death. As I walked home in the early morning, I asked myself, 'What am I doing in that environment?' I could not shake of Tim's accident, I could not eat proper food and it did not help when my mother said, 'We've got a lot to thank God for. Your father intended to put *you* as taker–off and put Tim on the waggoning for he's a bit older than thou art.'

Each day as I came downstairs to begin my daily study or to go to night school I felt unable to concentrate. My parents noticed my change of habits and the serious frame of mind I had fallen into.

'Where hast been to all the evening lad, what art thinking of all the time?' These questions were put to me almost daily and I could only reply, 'I don't know, I really don't know. Since Tim's accident I keep wondering why I have to go into the bowels of the earth merely to earn a living.'

My father was sympathetic, but his only remark was, 'Thou't get used to such things happening in th' pit as time goes on.'

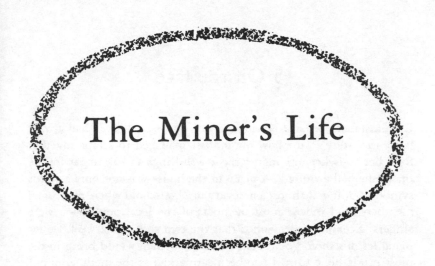

The Miner's Life

5 On the face

Circumstances began to change in our domestic life. It had always been a mystery to me how my mother managed to scrape together from her housekeeping money the few shillings a week to pay for the piano she had acquired. A piano in the house was not only a status symbol, but it was almost a necessity in a household where most of its members, and indeed most members of the local population were singers. Economics demanded that the two shillings a week fee for piano lessons should be invested in a way which would bring in the most profitable return. I had been earmarked as the member of our family who would show some return for the investment in the shortest possible time.

Over and over again I asked myself, 'How does my mother find the money for the hire–purchase?'

Perhaps my wages coming into the house had made the investment possible or perhaps my father had been given a rise in wages. I never knew quite what it was, but the news came suddenly that a piano was coming into the house. It was not long after this event that other changes were made, for we moved from Abbey Street to a new address in Park Road. This was one of a short terrace of ten houses at the very end of Park Road which was situated on the pleasant outer edge of the village.

Secretly I suspected that it was a little too ambitious of my parents to move for I still retained vivid memories of near starvation during strikes in my early childhood. Now, the echo of these fears was accentuated as I realised that the piano and the move to a more pleasant environment would amount to an increased burden on the family purse. My mind queried . . . 'Can we afford it?'—Such had been the deep awareness during growing–up years for the need to watch every penny. I knew that my parents had no money saved up and that we lived from hand to mouth; every penny of wages was mortgaged each week. But I enjoyed watching the pride and pleasure in my mother's countenance once we had settled in this slightly more pleasant part of the village. We were all warned to be on our best

Victoria Street, Silverdale, Staffordshire. The author was born in this street.

A few miles away is Keele Hall, within which estate were located the coal mines of this book. It was the home of the Sneyd family, who received one shilling for every ton of coal extracted. It is now part of the University of Keele.

Pitboys with two young pitmen. The length of the pit-props stacked on the pit-bank indicates the height of the seam.

Face-workers on a 7–8 feet seam. These coal-face conditions would be regarded as 'comfortable' on account of the headroom. Note the drawers worn by one of the colliers.

A 'heading', a new road, being driven in. The solid wall of rock and the 'bridge-rails' provide a temporary track until the more permanent rails are laid.

An excellent photograph of a rip. The size and height of the old road are indicated by the arms of the two rippers. The clothes of the men suggest that it is not a hot place.

A small area of the district called The Potteries, where the author grew up. The congestion of the potbanks and other industries is marked in the photograph. This atmosphere was quite normal during the days of coal-firing.

Fustion-mill workers at Silverdale, about 1920. See Chapter   . Note the 'cloak-room facilities'. Lighting was by candle.

Coal-pickers during the 1912 Strike. The author was six years of age at this time and suffered severe hunger, but he performed his stint on the tips, turning over every shovelful; not a scrap of coal escaped.

Well organised coal-picking during the long 1921 Strike. The sieve indicates the thoroughness of the search for scraps of coal; the sleeping miner suggests his long hours on the tip.

The ill fated Minnie Pit Halmerend. The sombre scene after 155 men and boys had been killed, in an instant, one morning.

Mossfield Colliery No. 1. Rescue Brigade. Every pit has its volunteer rescue-men. This photograph taken at the Rescue Training Centre shows the men wearing the face-mask. The oxygen is drawn from the large bag suspended from the shoulders.

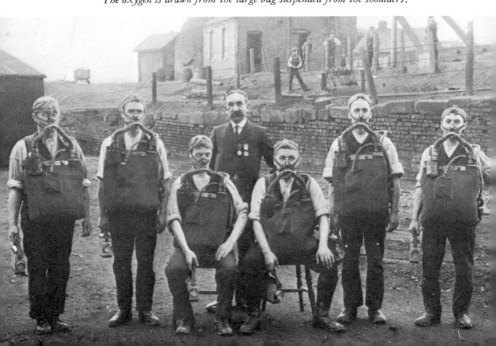

The pit referred to in the book as the Number Six. The author's father and brothers worked at this pit; it was a hard place to earn a living.

Kents Lane Pit, Shafts Nos. 14 and 15. The unusual tandem headgear, now used as upcast shafts, is noteworthy. These were the shafts used by the author in his early pit-days. The thickly wooded surroundings support the author's description.

Silverdale Colliery as it is today. This shaft was sunk to a lower level and widened considerably about the year 1923. The author used this shaft for about two years; it stands a quarter of a mile from the twin-shafts Nos. 14 and 15.

The author's son (right) with Mr. B. Griffiths the Training officer at Silverdale Colliery after he had escorted them on the tour to the coal-face described in the Epilogue of the book.

behaviour; we were proud of our home which was a credit to my mother's housekeeping and my father's provision. We were regarded in the village as a strong chapel family and we did not owe a penny to any trader. Our outstanding liabilities were the hire–purchase agreement for the piano and the mortgage on the house which my parents had purchased for the sum of £220. Before making this daring investment they had paid rent over many years to the same local builder who had given them the first opportunity to buy this house. Only my mother knew what sacrifices had been made to acquire the piano and now the improvement in our standard of living; we, in turn, knew the pleasure we all enjoyed because of it. Also, we were now out of earshot of the THUD . . . THUD . . . THUD . . . of the great steam hammers at Knutton Forge.

At night when the wind was in the right direction I could hear the express trains hurtling along the main line from Manchester to London as I lay awake in bed. This experience captured my youthful imagination; the clattering of the wheels as the trains sped on over the straight stretch of line filled me with curiosity and romantic ideas. It gave my imagination great play and filled me with hope that one day I might escape from the pit. The sound of those trains made me ask myself, 'What course can I take so that I can get out of the imprisonment underground, escape from the trap I am in and become a passenger on one of those express trains?' I had become aware that I was living through the same monotonous routine which my elder brothers had experienced; arriving from the pit each day, washing my hands, having a meal, resting my head on the edge of the table to sleep off the top layer of the sheer exhaustion. Waking again as my father arrived from the pit, another cup of tea, washing off pit–dirt and changing into cleaner clothes, then off to night school or retiring to the cold parlour for piano practice and private study. What a routine it was, exactly as I had witnessed with my brothers as I had grown up from early childhood. I felt that I was now caught up in the same soulless procedure with nothing to look forward to in life but . . . THE PIT.

In spite of these occasional pessimistic periods of reverie, I applied myself diligently to the study of mining by attending evening classes and doing private study. At the same time I paid serious attention to the playing of the piano, although working in the pit had played havoc with my fingers. With the heavy work the dirt and the grease became embedded in my flesh hampering my technical progress on the keyboard. But my music teacher had assured my father that he was not wasting his two shillings a week and that I was making good progress. This was proved by the Trinity College of Music examinations I passed from time to time. I usually passed with high marks; indeed in 1923 I won the prize for the highest marks gained in one particular class. But to all intents and purposes, I was *still* a prospective candidate for the examination which would qualify me as a pit–official and, later on, much later on, as a colliery manager.

Now came the day which I knew would have to come; I was taken away from Bill Fry and Johnny Cameron on the waggoning length. My father announced. 'You will be all days from next Monday and you'll be working on the face up Two's dip.'

I wondered why I was being put on the face at such an early age; I was not aware that the Colliery Manager was behind these moves. But isn't working on the face a man's job?' I asked.

'Ah, Ah, it *is* a man's job. Thou hast asked me this kind of question before. Workin' on th' face is a man's job and thou't a man.' With that remark the subject was closed and I faced yet another new experience of work underground.

I sat at the bottom of Two's dip and waited. I kept thinking of Tim Jones and all that had happened further along the road at Six's dip. I was waiting for the arrival of Mr Abe Deakin, a man I knew only slightly by sight. I had heard my father talk about him, always with glowing compliments about his standard of work. I had noticed that he was among those who went down the pit early so as to be ready to start work on the stroke of seven. As he approached me I stood up and was about to speak, but he raised his lamp to my face and said kindly, 'Hello young man, I expect that you are Mr Brown's boy?' What a pleasure this was for me; the sound of the soft voice. The polite attitude, no swearing, no familiarity with my father's Christian name. He had said, 'Mr Brown's boy', not 'Art thee George's lad then?' as Cleer had said. This was a small thing, but it was important to me and it pleased me; already I liked Abe Deakin.

I looked up to the experienced miner and replied, 'Yes Mr Deakin, I am Harold Brown. I am to start working with you, I hope I shall be alright.'

'Oh, you'll be alright Harold, my son. You *are* rather young but being small will be an advantage on this four feet six seam. I will put you right, so don't worry lad. Come on, let's get up the dip; it is very steep so watch how you go.'

We were both panting as we reached the jig–landing at the top; I stood and gazed at the black crackling seam of coal.

'Here Harold, put your shirt down here with mine. By the way, have you brought any drawers?'

'No Mr Deakin, I have no drawers, but I expect my mother will get me some'. I found it easy to converse with this man; he did not use the colloquial way of speaking and I found that I did not use it in my replies to his questions. There were others like him in the pit, but such men were rare. As I became more deeply involved with him in our pit–work I always regarded him and his kind as men who would have made good statesmen or administrators. Already in this first half–hour of his company I was attracted to his good clean English and his ability to get along without the use of our own unique mode of speech, attractive as it was when not overdone.

Already I knew that there would be nothing for me on the face but hard work, but I knew that I was going to enjoy having this man at my side. Deakin approached me, standing only in his clogs and drawers which resembled the bottom half of pyjamas. I was fascinated by the sight of the coal face, but sweat was oozing from me even as I stood there, doing nothing. Deakin smiled and said, 'Now, here you are, here's your box and rake.' The box resembled a dustpan in shape, but was made of strong iron or steel and its capacity was that of about three shovelfulls of coal; the rake was shaped like a garden rake, but it was bigger and stronger, made of the same material as the heavy box, really tough metal. It had a short handle.

Abe Deakin rested the back of the box on the front of his clogs and gave me a display of all the movements necessary to fill the box with only two plunges of the rake into the coal, coal which had been blown down and dressed from the face by the previous night–shift. Patiently he schooled me in this technique of hand–loading until I could be left alone, learning from trial and error. I repeated the words he had drilled into me, 'First look round at the space where your coal is to go, just as a man firing a rifle looks through the sights before he fires.'

Generously Deakin came over to me every ten minutes or so and gave me instructions, but I suspected that he was making an excuse to make it necessary for me to stop for a moment or two and so give me a very brief rest. Well before snapping–time my body let me know that I

was performing an exercise to which my muscles were not accustomed, muscles which hitherto had not been called upon to serve; I
was now becoming aware of their existence. Arms, shoulders,
back–muscles were in full use and I now could see the necessity for the
very short handle on the rake. Childhood memories came rushing into
my consciousness as I recalled the expressions of my elder brothers as I
sat on my stool by the fire listening to their daily complaints about the
heat, sweat and slavery at the old Number Six Pit. I recollected clearly
that during conversation they mentioned the 'nine o'clock back–
ache'. They used to vow that once the loader was over that, then the
body settled down and fell into a rhythm of its own and the toil
became slightly easier. Sweat was now running down the whole of my
body and the nine o'clock back–ache had made its appearance rather
earlier for me. The stench of sweat from the other bodies all within a
short distance of me and within a very confined space, the other
nauseating smell which I did not yet understand, all this made me
wonder if I was going to stand up to these conditions. Two things
gave me the moral strength to fight on; I knew that what I was doing
was the kind of work my brothers had done up at Number Six, and
what they could do I considered I ought to be able to do; secondly, I
had Abe Deakin and his kindness. He had taken trouble to teach me
rudiments in loading and he had addressed me as 'Harold, my son'.
The reader may murmur the word 'sentimental' on reading those
three words, but will not understand or appreciate the almost
complete absence in our family of the quality of endearment and
affection. This is not a criticism of my parents; we were a large family
fighting for survival and any expression of affection was a luxury we
could not afford as our parents fought and struggled to clothe and feed
us.

At the time I was unaware that the coalface represented the might
of Mother Nature and that she was my opponent. As I summoned
every bit of strength in my boyish body and every element in my will
to do battle with such a contender I enjoyed a glow of companionship
which I had not experienced before; it was the realisation that this
older miner was taking trouble over me personally. At first I had
wondered why I had been placed with him and not with one of the
rough diamonds of the coalface. Before the end of the shift I knew that
Abe Deakin was neither soft nor weak. He was gentle in manner and
human in approach, but he possessed all the qualities of a first–class

pitman. When the toil was almost too heavy to bear, I went down on to one knee, glanced across to him and entertained the comforting thought, 'At least I have *him* and that is worth all this work.'

Snapping–time came and I threw myself down flat on to my back and had to rest before I could eat. The others made fun of my utter exhaustion for they had all been through it. The smell of sweat from their bodies again nauseated me for it did not go with the eating of food.

Abe Deakin came to my side as I began to let my water run from my bottle down into my open throat. 'Hold on, hold on, Harold, steady with that water. You'll need some of that precious stuff for the second half. Always just sip a little water at a time, my son, if you can control yourself sufficiently.' Immediately I recalled my father's final words of advice, 'Keep away from thy water if thou cans't.'

I opened the parcel of sandwiches wrapped in newspaper and Deakin was amused at my careful examination. 'Well, young man, any visitors been to your food?'

'No, it seems to be alright. If they've been, they have gone'.

'And how's your back lad?

'Breaking in two thank you Mr Deakin, but my, it is hard going'.

'You'll get used to it. By the way, your father tells me that you play the piano.'

'I'm learning, but I'm not much good yet. Pit–work's no good for piano fingers. Are you interested in music Mr Deakin?'

'Yes, I play the 'cello a little. It makes it possible for one to express finer feelings and I think that the 'cello is a beautiful instrument for one to display these inner, intimate feelings.'

Now I knew *why* my father had placed me with this miner. Abe Deakin continued talking as we munched our sandwiches. 'You see, Harold, when you are doubled up here for seven hours a day with nothing but darkness and nasty smells, you can go home, get out your instrument, close your eyes and enter another world with music. One must be able to *express* oneself and shut out the blackness of this life at the coalface.'

'Yes, Mr Deakin, I know what you mean, that is why the miners are so enthusiastic about singing; it is their only means of balance. Without it, or without some means of expression as you put it, they would go mad working as they do under such pressure and under such horrible conditions.'

'My word, lad, you do seem to have thought it all out.'

'Mr Deakin, what do you do when you want to go to the lavatory?' In the shunt we were able to go to the pan which was just through the air–door leading to the air–road. During my waggoning on nights I never needed it, but I feel that I shall have to use some sort of convenience up here at the face. What do you do?'

The older miner laughed heartily at my inexperience, but it was a friendly laugh. He hesitated before speaking again.

'Well now, Harold, that pan you mention is the *only* one I know of in this district. It is too far to go from here, even if you used all your snapping–time for the purpose; you would not get back here before drawing of coal recommenced and you are not allowed on the road during coal–drawing.'

'Well, what *do* you do when you want to answer the call?'

'We have to use the gob. You see here lad, the gob is on both sides of this jig–dip and since it is being filled with dirt, it does not matter what men do in the gob itself. You will have to be careful where you tread because others will have been in there and the surface is not covered with dirt during every shift. This is one of the disadvantages of working underground; we have to put up with such things.'

'Phew so *that's* what the other smell is then. I wondered. Mr Deakin, what exactly is the gob?'

'Briefly, it is the space which is left after the coal has been extracted. As the seam is hewn and blown down and the coal sent to the surface, quite naturally there is a wide area of unsupported roof. It is impossible to maintain support of such a wide area of roof, that is why some refer to the gob as 'the waste'. The sides of the jjg–dip are built up with stones or other solid material, forming a rough wall. Any rubbish can go into the waste area of the gob, but there's a strict rule that no timber or coal must be left in the gob. That would be regarded as combustible material and when the weight of the roof comes down causing extreme pressure it might cause the combustible material to ignite and create a gob fire. This is known as internal spontaneous combustion. Snapping–time is up so you must ask your father to explain the danger of timber going into the gob. There is quite a lot to it and he has had a great deal of experience in such matters.'

There had been little conversation from the other face workers during snapping–time. Once they had eaten their sandwiches, they lay back on the ground; all apeared to be utterly exhausted.

Deakin took my arm and led me to the face. 'Look, Harold, take notice of the distance from one post to the next one. Also how far the coalface is from its nearest post. As we fire shots and blow the coal down, so we strip off the face and move forward, further into the seam, ready for the next shot-holes to be drilled. The important point I am trying to make, Harold, is, that before we can move forward even one yard, the timber has to be moved up to obey the strict rules, to conform with regulations. When the face has moved away too far from the rails carrying your waggons then the rails themselves have to be moved forward, as near to the face as possible. When this happens we call it, 'shifting up' and this happens every few days according to what kind of coal we are working in, hard or soft. It varies you know from day to day. Each time we shift up, the jig–dip becomes that much longer and the seam that much further from the main road.'

'Thank you, I am beginning to recognise the pattern now. Every yard of coal taken from the seam involves that much extra distance in the transport of the coal to the surface.'

'Well done, you have it exactly; imagine that on a face nearly three quarters of a mile long, then multiply that by the number of faces in each pit. Think of the cost of merely shifting up in one pit alone.'

I moved as if to begin loading again, but another young man was already loading at my point on the face. No–one tried to disturb Deakin as he instructed me. I felt a glow of gratitude as I realised that I was receiving personal tuition from this expert. Suddenly Deakin took my arm again. 'Talking about it, Harold, won't earn a living. Come and try your hand at drilling a hole. Ever used a drill before?'

'Oh yes, I drilled for Mr Cleer on a rip in the air–road.'

'Ah, lad, that would be a ratchet–drill, a ratchet–machine.'

'Yes it was, it was very heavy.'

'This machine is not so heavy and you need no post. In fact with this low roof it would hardly be practicable. Here let me show you; place the machine on your shoulder, using this old piece of cloth for a pad. Turn the handle of the machine right round as though you were turning a wheel in full circle. It takes patience until you feel the drill beginning to bite into the coal. Watch carefully now the way I do it.'

I *did* watch carefully, but my mind was on the words which recurred in this experienced miner's instructions to a young lad, sent up here to the coalface to learn what methods were used to extract the coal. He had said, 'Here, let me *show* you,' 'It takes a bit of *patience*,' 'Watch how I turn the handle *gently*.' I knew within me that Abe Deakin had decided to give me instruction on this other aspect of work on the face merely to provide me with a break from the arduous, back–breaking work of loading. Even as a young lad, on my first day with this man, I recognised his sterling qualities. My back was breaking from the morning with shovel and box and rake, and Deakin knew it. He did not just leave me to get on with the drilling, he stood over me, told me to 'sprag' out my legs a bit in order to bring my body in direct line with the hole to be drilled. He never hurried me nor gave me the impression that I was wasting his valuable time. He accentuated vital elements of the technique. 'Harold, keep your head "a-one-side" and never let your eyes leave the end of the drill where it enters the coal.' To a man on the face, this was all in a day's work, routine procedure; to me it was a wonderful experience on account of the quality of the instruction and the character of my teacher.

The position I was in to do this drilling was taxing even more muscles of my body and I began to wonder if there were any muscles left within my frame which had not been taxed to the limit since I entered this world of slavery and sweat. My advantage here on this seam was that I was small. At last I felt the drill grinding into the coal and, although my shoulder ached with the strain, I now felt confident and full of satisfaction. When the whole length of the drill was buried in the coalface I made some attempt to wipe the sweat from my face with a very black hand; I looked around and watched for a moment the other black bodies going at it for all they were worth and the thought came to me. 'How do they stick it, every working day of their lives?' Deakin returned to my side when he saw that my first hole was finished; he gave me a warm smile as he withdrew the drill and said, 'How about another hole now on the far side?' Mr Bassett will show you where to put it; wait till he has finished his holing at the base of the seam. That space which he is cutting with his pick provides an area of least resistance so that when the shot is fired the coal will be blown downwards, see?'

He had told me to 'WAIT' till Mr Bassett had finished his holing. What a wonderful word that was 'WAIT'; I felt that I wanted to vow never again to take for granted the privilege of being able to relax, even for a moment or two. Bassett, seeing me with the machine in my hands, got up and pointed to the spot where he wanted his hole to be drilled. Here was another of the many men whom I knew only slightly by sight. He smiled as he wiped the sweat from his face with his bare, dirty hands. 'How't gettin' on, Harold; how wouldst like to be back at skew?' The mention of school ('Skew') brought a flash of pleasure; I went down on to both knees in order to have the ease of straightening my back. 'Oh dear, Mr Bassett, school you say; what a lovely life that was, compared with this black existence.' He merely smiled again as I put the piece of cloth to my shoulder, aimed my drill and began again; my second shot—hole of my experience at the coalface.

As I buried my drill deep into the coal, Bassett called across to Abe Deakin, 'Eh, look thee 'ere, Abe, wey'n got another collier on th' face. He drills a hole like a mon, what hast been telling him then? I reckon that thou hast been giving him private lessons.' That was the language I wanted to hear and whilst it fed fuel to my boyish vanity, Bassett's praise warmed me as much for the sake of Deakin as it did for my own satisfaction. Secretly I sensed a feeling of personal pride and my father's pleasure in good workmanship crossed my mind. I stopped to wipe the sweat away which was running down into my eyes. I took a look along the face at the other men, all hard at work and inwardly I said to myself, 'What fine men they are, if only people on the surface could see them now.'

The fireman came to fire the shots, *my* shots; I had taken a hand in this firing. I wondered why my father was not doing the firing, but I learned that he had gone to the surface to attend a lecture on Rescue Work. I wanted to see every aspect of this firing on the face. I saw the fireman take Deakin aside and whisper a few words; then they both looked across in my direction and I knew that I was the object of their conversation. The fireman, a young man recently qualified, smiled at me in a kindly way and he motioned to everyone that he was ready. 'Ready . . .' he shouted as he connected the detonator—leads to his long flex. The men dispersed along the face, the jigger went part of the way down his dip; all approaches to this part of the face were now guarded. Then came that final moment; the fireman gave his last look round before connecting the flex to the firing—battery. There was a moment of silence and my father's words to me flashed through my mind for I remember once at home he had been relating a moment when he realised that he had a misfire on his hands. He had said to all sitting at the table at home, 'No—one will ever know how many prayers are silently uttered each day down in the pits of the world; how many firemen send out a little private thought that there will be no misfire and no disaster because of this shot.'

'Firing,' shouted the fireman. Then came the loud explosion and the *dust*. There was nothing to do but to go on breathing it down into our lungs; there was nothing else to breathe in. Everyone coughed and sneezed, then spat out the muck from their throats and again I said to myself, 'Here is that dreaded dust again.' The dust at the coalface which always seemed to be there so nauseated me that it created an aversion to dust of any kind which has remained with me throughout life. We remained strictly away from the face while the fireman tested the roof with the end of a pick–shaft. He then gave instructions regarding timber and left us. As I returned to begin loading I overheard one of the colliers remark to his mate, 'Didst notice, Jack, nothing escaped him. See how he spotted that post, only six inches out, but he made me shift it and put it in line. This young one never stops asking th' owd ones questions.' I recollected that I had seen this man with my father at home during his period of preparation for his fireman's papers and I knew that he respected my father's ability.

The arduous shift came to an end at last. The colliers made little comment, but quietly put on their clothes and left. Changing from his drawers into trousers Abe Deakin watched me drain the last drop of water from my bottle, for I must have lost two pints of sweat that day. He came and stood over me, 'Well Harold, how do you feel? You're tired I expect. You have used muscles which normally you do not use and so you will be a bit stiff tonight.'

I rose to my feet and replied, 'Well, Mr Deakin, I was stiff enough after my first day in the pit and it can't be any worse' than that.'

He laughed, Oh yes it can, you wait till tomorrow morning, my son. You will feel it under your crutch; working in this height makes you obliged to sprag out with your feet and legs to bring your height down. Not so bad perhaps for one your height, but you have had your legs apart for most part of the day. But you have done well for your first day. Tomorrow I will show you something about how to set timber on the face. Are you going to hurry off up to the pit lift like the other young men who race off at the end of each shift?'

'Hurry off, race up that dip? Mr Deakin, I'm too tired even to walk it, let alone hurry.'

'Alright then, you won't mind my company for I always take it at a nice steady pace. We can continue our talk on music.'

'I would like to thank you, Mr Deakin, for the trouble you have taken to teach me things today.' I voiced my gratitude in a timid sort of way, but my thanks were as genuine as it was possible for them to be.

'You are a good pupil, Harold, and there is nothing I would not do for your father, you know. I've worked for him for many years.'

Secretly I was proud to be seen walking out of the pit with this experienced miner. A man who had learned a difficult job the hard way and, in spite of his skill and knowledge in this arduous life, he remained modest, quiet and gentlemanly in all he said and did. As we walked up the main dip he explained to me that the present methods of getting coal would soon be at an end and that huge cutting machines would work the whole length of the face and replace the many small groups of colliers.

'Have you noticed these pipes laid alongside here on this main road, Harold? Very soon they will convey compressed air to the face from the air–compressor on the surface. Then you will hear the roar of the machine instead of the ring of the pick.'

Then our conversation turned to music and I was amazed at the extent of his knowledge of some of the great composers and their works. I felt honoured to be sharing this conversation and my father's words again rung in my ears, 'When Mr Deakin is addressing you, keep thy tongue still and listen, dunna interrupt him now. Take notice of all he says to you. And watch thy manners mind.' I really did not need this warning for I found myself hanging onto every word which Abe Deakin uttered. His quiet manner, clean speech, absence of swear–words, his conscientious workmanship, his control; all this stimulated my admiration for him, *and* he was an expert miner. I wondered how many people in our village even knew that this quiet citizen played the 'cello? There would be even less who were aware of his knowledge of Beethoven, Bach, Mozart and Handel.

That day after my meal I felt more that ever ready to let my head rest on my arms on the edge of our living–room table. Sleep took over in a second and my father eating his meal did not wake me. When I did rouse myself my father was reading the paper and the table was cleared. He looked up, 'Hast learned anything today, Siree?'

I did not reply for a moment for I was not sure in what realm of learning my father's question was meant to be. Had I learned anything about pit–work or about Abe Deakin? 'Learned anything? Learned anything?' I said softly, trying to avoid my father's eyes. 'Yes I have learned something; I now know what the nine o'clock back ache is. I know what it is makes loaders champion swearers in the pit. I have drilled a couple of holes into the face and I expect my shoulder to drop away from my body at any moment. I have learned what it is like to choke with dust in my throat; I know what it is like to work with other men in a confined space as hot as hell. I know what the smell of their sweating bodies is like plus the stink from the gob after the men have been in there and done their business. *What* an atmosphere. And you ask me if I have learned anything. The coalface can hardly be compared with the halls of learning can it?' My father kept his eyes steadily fixed on me, not quite sure whether or not I was deliberately being impertinent, while I did my best to conceal my facetious frame of mind.

'Several new skills have been added to my ability as an apprentice pitman and I suppose I *have* learned quite a lot for one single day at the face. But one thing stands out above all that I have just mentioned, something which has surprised me and which has made all my effort worthwhile.'

My father dropped his newspaper to the floor and again fixed his eyes on me. With a crescendo he ejaculated, 'Well, go on lad, go on, what hast learned then?'

I suspected that he did not know what to expect from me. Some young men, once they are working on the face assume the stature of a big man and tend to become a little over–assured. I could see from his manner that my father anticipated something impudent. But impudence was the very antithesis of my mood at that moment. I searched my mind for suitable words. 'I have worked with Mr Deakin at the face; my very first day and he has gone out of his way to teach me a good deal about the standard expected of a loader. He has introduced me to other aspects of work at the face and how I should tackle certain difficulties. He advised me to talk to you about one or two things. But he has taught me something much more important than the business of getting coal from the seam and sending it to the surface. He has shown me what a *real gentlemen* is like.'

My father took up his paper again but his gaze was in the corner of the room; something in the far, far distance had captured his mind, his countenance had changed. I could see from the small white patches which had appeared at the top of his cheek–bones that emotion was taking a grip of him. He looked across at me. The word 'pride' would have been an essential part of any interpretation of that look. I was happy that the words I had chosen to describe Abe Deakin had given him pleasure. It was not often that my father afforded himself this kind of luxury for, in spite of his strictness, he was a modest man. Impatiently he urged, 'Well, go on. Tell me what has happened at th' face.'

I was obliged to hesitate before trying to translate my inner pleasure into simple words. 'I did not hear Mr Deakin swear once throughout the shift; he does not use our local way of speaking; you know what I mean, I wunna, dunna, shanna, costna etc. He uses the proper words instead of our North Staffordshire way of saying things. He always speaks quietly and clearly as though he is living in another world. He seems to be so much in control of his temper, even though other men around him are swearing; he just does his work quietly and treats everyone as though they are a special person. At Sunday School all those teachers used to tell us that if we controlled ourselves we could live like that and be an example. Older lads used to say that you could not do that down the pits, but I have *seen* it today, a man living and working just like that. What a nice man he is, and he is a musician too, he plays the 'cello and knows a lot about composers. He is a real gentleman working under dirty and difficult conditions with no comfort and no facilities for decency and cleanliness. No means of washing one's hands before eating food. The worst conditions ever; even to go to the lavatory means going into the gob and behaving like an animal. There he was, a gentleman, and I hope that I can be like that one day. One thing I noticed in particular, although the men swore like troopers, they never used bad language when speaking to Mr Deakin; it was as though they respected his way of living.'

I now had no doubt that my father had put me with Abe Deakin because of his regard for him as a good worker and a good collier but the fact that Deakin was keenly interested in music must also have been a contributory factor. Not only had I returned home that day having been introduced to a fresh field in coal–mining with the attendant new skills, but I had returned with my mind enriched. An element of lustre had come into my life of drudgery underground, I had found a man who knew how to live properly, self–controlled and I knew that he was willing to share his knowledge with me. My father listened attentively. I saw the rims of his eyes become red and I knew that he had been touched by the things I had related to him. He swallowed with some difficulty and said, 'I'm glad that thou't getting on alright with Abe. He will teach thee th' proper way to do things, so take notice of all he says to thee.' My father rose and went to the door, but he turned and, after a moment's pause said, 'Oh, and look thee here, lad, make sure you give him the respect due.' Then, as if the result of an afterthought, he asked, 'And how didst get on with thy loading?'

'Loading is very heavy work, especially with box and rake, but I expect that I shall get used to it in time.' My father then left the room, but I could see that he was deep in thought. As he slowly closed the door he gave me one last fleeting glance. I tried to imagine what it was that made him do this and I was convinced that there were many qualities contained in the glance. To separate those qualities would have been like trying to separate the colours in the rainbow. Whatever the constituents of the fleeting glance were, I felt sure that the quality of 'pleasure' was the dominant one.

The days wore on and each day that the fireman came round to fire, my interest was stimulated, for what he did was the very substance of my study at night school. Sometimes it was my father who fired the shots and made sure that I was at his side as he fired. To me this was the most important event of the day for I was able to participate in the vital function of turning down the flame on my father's lamp as he tested for gas. I was able to see the practical application of the theory contained in my study from books and instruction from teachers. I watched each time I saw a collier setting timber and, more than anything else, I found that my ear was becoming sensitive to noises up in the roof. I began to recognise the harmless goth explosions above my head and then the more ominous rumblings which demanded . . . 'Silence everyone, listen, keep perfectly still.'

I took my pattern from Abe Deakin for it seemed that he had his ear on the roof every moment of the shift. Often he would listen to the roof with some apprehension and alertness and then, as if by some inner instinct similar to that possessed by birds, animals and insects, everyone stopped and there was deathly silence until Abe moved and all was clear again. I tried to copy Abe Deakin in every respect during the shift's work. When things became difficult and stress built up, I tried to remain calm and speak with a low but clear voice with my mind on one criterion, 'What would Abe Deakin do?'

I found it difficult to become accustomed to the obnoxious smells of this life on the face; heat, sweating bodies, smells from the gob; it all nauseated me. Like other men I had to use the gob for a lavatory and I would have given anything for the luxury of being able to wash my hands before my snapping. Such were the conditions at the coalface, such were the discomfort and the danger, and all for a pittance of a wage. I was taking home less than two pounds a week. Six days of hard work for such a small amount and wages were now in decline for the spectre of industrial collapse was on the horizon.

A close bond of friendship grew up between Abe Deakin and me. We always ate our snapping together; we walked out of the pit together and we seemed never to have enough time to discuss all our common interests. All that Abe Deakin's set of men did and said at the coalface made a deep impression on my mind. The courage, the guts necessary to perform an ordinary day's work were not taken for granted by me. They faced their difficulties with cheerfulness and they seemed to possess an inborn philosophy regarding their arduous work, similar to the attitude of deep—sea fishermen and those who give service to mankind on the lifeboats. They accepted me as one of them and I never heard a word of insult between man and man. I am sure that the ever awareness of danger made insult to each other impossible. I cultivated a deep respect for these men who faced danger and such appalling conditions so that their families could have the necessities of life. The presence of Abe Deakin at my side was an inspiration and his conversation gave me immense pleasure. Sterling standards were being formed in my mind, standards by which I should set my values and opinions for the rest of my life. I was witnessing an example in 'manliness' and it was by this example that I came to rate all men that I met in my future career.

This admiration for miners, this attitude which I was beginning to cherish, was accentuated and strengthened even more when one day an incident occurred which most miners encounter perhaps only once in a lifetime. It had been a normal morning at the face, non—stop loading and drilling of holes into the coal. We all had withdrawn from our work and sat together to eat our snapping. We sat just a yard to two inside the jig—dip because the air there seemed to be slightly cooler. We arranged ourselves on opposite sides of the rails which took the waggons on to the landing. The general conversation was centred upon our local football team and its progress in the Cup competition. SUDDENLY, everyone stopped talking and there was a deathly silence for a split second. Bassett who was sitting nearest to the face shouted, 'Eh, jump, jump.' I felt myself pushed headlong down the steep dip. Deakin had given me a push to try and throw me clear for he, as usual, sat next above me. As Bassett had yelled there came a loud 'THUD' and the whole place was filled with dust. We were in complete darkness for our lurch downhill had knocked over the lamps. Recovering from the initial shock I felt a body lying across me. There was silence for a few seconds, the dust had made speech impossible.

'Bloody hell, what's up?' a weak voice enquired. The came the sound of another voice, 'What the hell was it? We are all inth' dark.'

Then came the quiet voice of Abe Deakin. 'You alright Harold?'

'Yes Mr Deakin, I'm alright except for my leg which hurts. What about you, are you hurt at all?'

Before he could reply another man said, 'We had better just stay till we get a light.'

'Yes,' said Deakin now with more firmness in his voice, 'Just stay as you are lads, don't move, there has been a slip.'

As he said this a light appeared at the dip–bottom and someone called out in the usual manner of pit–language, 'Dost hear, dost hear, anybody there?'

'Ah, we con hear thee, come up here with thy light', shouted back one of our group.

The slip had been heard along the face and men were enquiring what the situation was. Slowly we dragged ourselves to the bottom of the dip and out on to the main road. My memory has not served me well enough on this matter for me to be able to relate in detail what the injuries were for I was still in a state of shock. Several did suffer slight injuries and Abe Deakin was off work with his back. I had a leg injury and was trembling still when arrived at pit–bottom. On the surface I was given hot tea and ordered to get off to bed when I arrived home. My mother did not fuss over me for it was obvious that my injury was not serious and she had, on more than one occasion, received other members of our family arriving home from the pit with injuries. I told her as much as I could remember and her only remark was, 'Thank God it was snapping–time, thank God for that at least.'

I shuddered at the thought of what might have happened had the slip come ten minutes earlier for I had learned from my books, from conversation at home and many other miners that when a slip does occur it simply flattens out everything under it, like a pancake. The face collapses without warning and the roof comes down in one solid block. Ten minutes earlier we were all working under it and there would have been too much noise for us to have heard what it was Bassett had heard as he gave us all the split–second warning. Hearing my father return from the pit I left my bed for I could not sleep. Downstairs I went, for I had a dozen questions to put to my father on the subject of 'slip.'

My father sat quietly and I tried to assess his mood. 'What *is* a slip really? How often does that kind of thing occur? Is there *never* any warning?' Question after question, but my father quietly ate his food. At length he pushed away his plate and turned to my enquiring face. 'A slip, a slip. Ah lad, it rarely happens. It is something I hoped you would not experience, but, for some reason which is hard to explain, the whole face just caves in without warning. In my long experience up at th' owd Number Six I had a few, but this is the first I've heard of here in this pit. I anna been towd much about it yet for I was on another job today. One thing is certain, you were all very fortunate, you wouldna have had a chance, had you not been eating your snapping. Monty Bassett has good hearing, due no doubt to his playing a musical instrument that requires a sensitive ear. Now perhaps you will see a bit of sense in the strict regulations regarding the setting of timber on the coalface.'

I related to him what I remembered of the occurrence and as I left him said, 'One element in the whole matter I shall not forget; it did me good to see it. I could not help noticing how those who came down from the other dips were so very concerned until they were assured that we all were accounted for.'

6 Tightening our belts

I had been much too preoccupied with study to pay attention to what was going on in the political world. During snapping–time and in the queue waiting for the cage, I heard grumblings and complaints about the decline in the coal industry and the failure of the unions to obtain what the men considered to be a fair standard of wages. Indeed, like the industry itself, wages were declining and pit–owners complained that they were losing money. 'Well', cried the miners, '*you* are the managers of the pits, why don't you modernise them? You made a lot of profit during the recent war. . .' But now I found that I was having to take notice, for silently into our midst was creeping a dimension which brought with it echoes of suffering during strike in my childhood. Vivid memories of those unhappy days came flooding back, memories of strikes which brought starvation. Now I heard a lot of talk about *strike* and whilst my studies kept my mind busy and off the unpleasant subject, young men raved about their unsatisfactory conditions and I observed that the countenances of the women and certain of the small traders were changing.

Nature has endowed animals and birds with means of protection against natural enemies; similarly, these hard–working people used their inherent cheerfulness and warmheartedness to combat their enemy; dirt, danger, accident, sudden death and bad conditions of work. But now I recognised the new enemy; it was, to my mind, the enemy of FEAR. Suddenly I had become aware of it again. Empty bellies and worried faces, women trying to manage on a pittance, late queues at the shops on Saturday nights to obtain the cheapest food. No work meant only one thing, 'no food', and that was hardship by anyone's reckoning. Tension mounted, vivid memories brought back with increasing regularity the humiliation of receiving a slice of bread and jam from the charity–box in infant school. Sometimes this would be all we should receive till the same slice was handed to us next morning by our teacher. It was now 1921 and the same dark picture loomed up before us all. No wonder countenances were changing.

We had been taught in Sunday School that 'Goodliness and

Godliness breaks forth into the countenance.' So, in the same way, I was observing that poverty and struggle to make ends meet are revealed in a person's eyes, shape of the mouth and depth of the lines on the brow. The short–lived and comparative prosperity of the 1914 war years had ironed out some of those lines, but now things were moving in the opposite direction. The deep furrows were again appearing on women's faces. My father, being an official, would be expected to perform certain limited safety duties, but he was a strong union man and he did not use his privilege merely to remain at work. I had great respect for his ideas and I retained recollections of his difficulties when making his decision on this matter during a strike. If he felt the strike to be a just one, he would join the ranks of the men fighting the battle.

Four years of futile waste in war had ruptured industry; no–one seemed able to convert the machine which had been used for destruction into a means by which goods could be produced for peaceful utility. There was no vision and mines and miners were the first to be affected by this lethargy. No–one took notice of iron–tipped clogs on the pavement during the hours when shifts were changing. But, let there be heard the ring of a solitary pair of clogs on the pavement during the working–hours of the shift and a dozen heads would appear in doorways and women would ask. . . 'What's up then? Has there been a fall of roof? Anyone hurt?' The reply would be returned, 'No, we have been knocked off, no trucks, no orders.'

'No trucks.' How often I heard that remark and I became worried for I knew from my studies on 'production' that 'no trucks' meant that no–one wanted to buy our coal. Short–time increased in regularity and I felt that I had been cheated. I had been sent down the pit because wages were better than shop wages. Now, instead of my usefulness earning me more money, I was beginning to experience gradual reduction. Every effort was made to avert the strike, but there came only deadlock and the strike was on.

The regular procession of men, boys and women to the dirt–tips was soon in evidence. Old cycles, baby–prams, hand–carts and garden barrows; anything in fact on wheels which would transport the bits of coal from the tips to the fire–grate. Within weeks the tips had literally been completely turned over. The weather was fine generally and no–one escaped their turn up at the tips; a few better–off local people generously *made* work around their houses to help large families in dire need. The most formidable enemy was hunger, and with it came an acute form of humiliation. People went out into the meadows and fields to look for pig—nuts, dandelion leaves or anything that could be used for sandwiches. Others put honesty aside and nocturnal excursions to the farm fields and farm yards were made for a swede or anything else edible. Some of the farmers, (bless their hearts) sometimes turned a blind eye to the petty thieving. The heartening feature of this degradation was the way a group of people made it their special task to beg a shovelful of coal—bits from each of those at the tips; this they distributed among the old and sick who were unable to go to the tips themselves. In the same way, they begged from all and sundry to make sure that these old people had at least a morsel of food each day. Such conditions brought the warm, human qualities to the surface. Pit—owners employed well—known poachers who knew all the tricks of stealth. These men guarded the stacks of timber which stood around the pit—bank, but, like some of the farmers, they too looked the other way when they knew that the bit of timber was going to a needy old person, or to a family whose need was acute.

After only a few weeks of strike the appeal for credit went out to the tradesmen. There was always a procession to the pawn shop even when there was no strike, but now this procession was in full swing and most of the unfortunate women knew that father's best clothes would not be claimed till this strike ended. The clothes would not be required for Chapel, for who could go to Chapel if there was nothing to put on the plate. As week followed week a good hot meal and the satisfying feeling it gives to a hungry man became a thing of the dim past for many, many people. My father's odd day of safety duty saved me from the depths of starvation. His exact attendance at the pit is something which my memory has failed to retain. But, regular meals as such had ceased. When once I complained, my mother turned and gave me a stern look and addressed me with sterner words, 'Hast forgotten that we are *all* fighting this together? Because thy father does an odd days work does not mean that his bit of money is used *only* for this hearth. Hast forgotten those empty bellies in Earl Street? Think of them with nothing coming into the house next time thou feelst like complaining'. There were plenty of empty bellies in Silverdale at this time. My mother sent me with a large jug of hot soup each day to a very poor family in Earl Street, a family where there was no father; this was only one case where my mother's compassion was too strong for her. The womenfolk were wonderful, their sacrifice too varied to be enumerated; Christian spirit carried out to the last letter.

Late one Saturday, standing among a large crowd on the open market site in the nearby town of Newcastle–under–Lyme, I noticed two of our day school teachers bidding for a large box of bones. Bargains were to be had at this late hour as stall–holders brought trading to a close. I wondered why teachers in regular employment should need this very cheap form of food, and it was natural for some of us to express our surprise. But the truth soon was out, for it was revealed that the teachers were part of a team engaged in buying up as many bones as possible in order to open a little soup–kitchen for the benefit of those children who had nothing in the way of hot food at home. Headmasters, teachers, Sunday School teachers, clergy and other volunteers went from door to door of those people not affected by the strike. They carried small tins and boxes, begging at each house for a spoonful of sugar, cocoa, coffee, tea, a slice of bread; anything in fact which could be used to give a child some satisfaction in the way of nourishment.

Despair took a grip of me amid this destitution and I asked myself, 'Why was I born into a large family?'

As stocks of coal ran out, factories, forges and furnaces closed and more men were thrown out of work. Then, gradually, a happier state of mind began to creep into our midst and ease the tension for I realised that certain benefits had come with this enforced idleness. The hardship had revealed and accentuated the excellent qualities of the people. Many bellies *were* empty, spirits were low, countenances were sad, features drawn, as we faced what looked like a hopeless future; but we were not broken. Men and women were settling down to constant hunger; the hardship revealed the potential of our community as warmth and comradeship were generated by the very struggle itself. The battle raged, but men and women began to enjoy their struggle. It was a wonderful feeling and those who believed in God said that His hand was upon us for we were now in the midst of the warmest and finest summer England had enjoyed for many years. We now had time on our hands which was not available when our slavelike occupations demanded all our time and energy. People walked around Keele and Finney Green to take into their lungs the clean, fresh air. Many could not take advantage of this for walking

101

demanded energy which, in turn, could not be generated without food. Wear and tear on clothes, shoes and clogs had to be considered. My two younger brothers and I were warned that we must wear clogs all the time now till the strike was over. We went to Sunday School only on alternate Sundays to keep us off our best clothes and shoes. But many miners wearing their rough pit–clothes did venture out into the country for the joy of clean, fresh air. Never before had miners taken to the lanes and fields in such numbers; the strike had provided the time and opportunity. Each day the morning summer sun beckoned me and on one such day my friend, Bernard Cliff, called and we could not resist the call of the clear sky and the open country. I wore my pit–clothes and clogs, but so did most other miners and no–one cared about it.

As we approached Holly Wood we heard the sound of male voices singing and we knew that already a group of miners had gathered to rehearse their male choir pieces and try to forget their hunger. Bernard stopped and we listened. 'Harold, how can they sing like that with empty bellies?'

'Bernard, you will not stop North Staffordshire folk singing even with their last breath. Has it not occurred to you that up here in this air, they have everything to make them sing?'

'Yes, Harold, except a good breakfast in their bellies. I've heard my father say that coming up from the dark dungeons underground into fresh air makes the miners keen to sing to express themselves.'

'Now I have heard *my* father say, Bernard, that the pit–ponies would sing if they could once they are turned loose into Park Field during Wakes Week.'

In Quarry Bank Road we looked right to admire the magnificent view of the Cheshire Plain and across to Snowdonia in the distance. Soon we were in Two Mile Lane, leaning on a farm gate and taking in the charming view of Keele Hall with its attractive sloping lawns and shapely cedar trees. The Hall itself gave us the impression of being a jewel set among the various greens and browns of nature. As we discussed the merits of the estate worker with his low wages, but with secure tenure, then compared him with the insecurity of the miners' lives, Bernard noticed that I pinched the buds of the hawthorn bushes and ate them omnivorously. He took the hint and brought out his sandwiches and at once we returned to Keele village, where Bernard entered the Sneyd Arms Hotel and brought out a bottle of lemonade.

Bernard was playing host for he was working. A Sunday School friendship was being forged into a strong bond of comradeship which lasted right through life.

As if by instinct we turned our feet towards Spring Bank and there again before us, spread out beneath this hill like a magnificent carpet, was the Cheshire Plain. Spring Bank stands alone, separated from the general range of hills which form an escarpment between the Cheshire lowland and the industrial Potteries of North Staffordshire, the numerous pits and the heavy steel and iron works. We both threw ourselves down on the lush green grass and took into our lungs the invigorating clean air which was blowing in from the distant Atlantic Ocean. After half an hour I could not contain myself. 'My . . . My! Bernard, there's one thing, no—one can take this away from us.'

'That's right Harold, but we don't get the time, do we? We are always at work and when not at work we are too tired to come up here. We should not be here now if you were not on strike.'

'It doesn't seem right Bernard. Here is this beauty, right on our doorstep. Now, because of your long shop hours you have no time left. I leave off work earlier than you, but my life is full of study and preparation for examinations.'

Between us, Bernard and I put the world right and I told him of a dream I had always cherished in my young heart and mind that one day I might become rich enough to build a hospital up on that hill where miners and pottery—workers suffering from the dreaded chest diseases caused by dust could be brought into that clean air. Even if they would never be cured, they could at least live their last days breathing in the invigorating air and look on to the magnificent view. 'If only we had this air down the pit, Bernard. The darkness makes day—shift just like nights; then there's the stink which gets right down on your stomach till you want to vomit. Oh if only you knew.'

'Ah Harold, but I *do* know. I've heard it from my father all my life. But up here there's everything to make us happy. Birds overhead singing, green grass, sturdy trees, buttercups, ladysmocks, scenery to delight our eyes and then all that yellow broom which covers these hills when it is out. All this beauty here and we feel so happy among it, yet two miles away our neighbours are near to starvation.'

Bernard insisted that we share what remained of his sandwiches and it was now approaching mid—afternoon. We had enjoyed the peace and quietness and we made as though to leave this pleasant hilltop

103

when Bernard suddenly took my arm and I assumed that he wished me to take a last look at the view along with him. But I was wrong; he said nothing for a few moments, then looking up at the trees said, 'Harold as we have been sitting here listening to the humming and singing of the wind up in the branches of the trees, I wondered who it was who planted them there in that attractive clump. They did it quite unaware of the pleasure people like us are now having, just because they took the trouble to plant trees. As the wind hummed its tune among the branches it thrilled me to think that not long before that very minute, the same wind had blown across the surface of the waves of the Atlantic Ocean, then rushed over this lowland plain and brushed the tops of these trees and given us pleasure.' I did not reply to Bernard's remarks for his little speech had gone down, deep into my mind. We sped downhill towards Silverdale, but we both were silent. I had never known my boyhood friend express himself in such romantic style. I always knew that we were born friends, but Bernard's last remarks had confirmed that there existed some deep spiritual affinity between us.

Our walk downhill brought us to that corner with an attractive name 'Crackley Gates' and soon we passed the well–known row of quaint cottages which for some reason had been given the comical name of 'Treacle Row'. Here we passed a man being led by a small boy; the man had his eyes bandaged and I saw ointment smeared on his nose and at the corners of his mouth. Bernard spoke to the man as he passed the time of day.

I was curious, 'Who is he Bernard? Is he blind?'

'I don't know him well, he's a friend of my father. He used to work at Apedale Furnaces, but now he works in the steel at Shelton Bar. I really don't know why his face is bandaged.' Silently I made a mental note for here was one more occupational hazard for me to investigate and enter on to my records. I felt sorry for the poor man being led by the boy for he was a man of robust stature.

My father sat alone before a small fire which evinced our economical stringency for even in summer we usually afforded the comfort of a good fire. He turned in his chair and asked me with some concern, 'Where hast been, lad, all the morning?'

'I've had a walk up to Keele and to Spring Bank with Bernard and the air is lovely up there. We met a man and a boy leading him. The man's eyes were bandaged; Bernard says that he works in the steel, what can be the matter with his eyes?'

'Anything can be the matter with a man working in steel just as it can with a man in th' pit; dust, nystagmus, broken back and God alone knows what other complaints which come from th' pit, the Potteries and the other heavy industry. There's lead poisoning in th' pottery trade and consumption in half a dozen other trades where there is dust. Why, look at th' men over in th' blast furnaces. they get all sorts of injuries, and imagine what it is like, working in that terrific heat and then, right out into the bitter cold wind of a winter's night. Perhaps the man you met had been associated with the lime in the steel work; I've heard it said that men get the lime into their eyes and irritation is set up anywhere on the face where there is moisture. Thou't understand more o' these things as thou getst older.'

I then told my father of my fantastic plan to build a hospital up on Spring Bank if ever I became rich. It would be for those with the dust, those who, because of serious injuries had been forced into retirement. 'Better to die up there in the fresh air with lovely surroundings than to rot away in their dark terraced houses with little sunshine coming in, old–fashioned sanitation and the flies that go with it in hot weather.' I had let my tongue run away with me and I watched my father's expression. 'Ah, lad, that sounds very good. A dream I should say, dostna think so? Harold, that is how it *should* be, but they'll never do things like that for workers. Also, what treatment wouldst thou advise for any miner with nystagmus? All the fresh air in the world wouldna relieve the pain of those with th' stag. Stag is semi–blindness brought on by years and years of straining their eyes under the glimmer of a Davy pit–lamp.'

I was grateful for this exchange of ideas for my father had not been over this kind of ground with me before. 'Well, I don't care what people might think or say about my ambition to have a hospital on Spring Bank and I must have mentioned it a thousand times, it would be a wonderful thing for miners and pottery workers. I am sure that I am right in this.'

With that remark the subject was closed for my mother called us to our meal. The middle of the day meal had been taken forward to the middle of the afternoon in order to bring it almost to the usual teatime. This so–called main meal was followed by a cup of tea and this arrangement eliminated the bread and jam which in ordinary circumstances accompanied the cup of tea at usual teatime. Also, by this arrangement of having the main meal so late, it made it less essential for us to eat anything else before going to bed. The management of food had become a fine art.

There was no inclination within me that evening to settle to study for my mind was full of the pleasure of the day and the significance of Bernard's remarks which had revealed a seriousness which I had not discovered before. I wandered down Back Lane and leaned on a gate, but soon I was joined by a man with an excellent tenor voice and a keen musician in general terms. He was one of that team who assisted with the begging of food for the hungry children and I admired him. As he listened to my praise of the public spirit which was manifest in our village at the present time he brought to my notice certain people who had distinguished themselves and brought credit to Silverdale. He mentioned Sir Joseph Cook and I recollected that this gentleman had been brought into our day school when I was a boy. My father often spoke about him. Joe Cook was a Silverdale boy who started at about the age of twelve in the same pit as the one where my father was employed. He supported his widowed mother until her death; then he married a local girl and emigrated to Lithgow in Australia where he became deeply involved in Trade Union activities and eventually in politics. Later he became Prime Minister of Australia and later returned to England as High Commissioner. Our conversation led us to talk of other men who, through personal effort and industry, had risen to distinguish themselves. One was a man who excelled in the study of oil production; he was honoured for his work in oil exploration in the Persian Gulf. My friend became very enthusiastic as he watched me making notes on his remarks. 'What a credit these people have been to our village, Harold, for they all had humble beginnings. There is one who is still among us: George Bromley, awarded the DCM for brave and distinguished conduct on the field of battle in France. He was also awarded the Belgian Croix de Guerre.'

'Yes,' I replied with pride, 'George Bromley also was brought into our classroom by our Headmaster, Frank Ellams; that was when the war was still on. One of his brothers was killed in the war, another one wounded and yet another one taken prisoner. So that family made its contribution to the war without question.'

So we talked and talked as we walked back towards our house. He seemed to have endless enthusiasm for those who brought credit to our community.

Next day it was my turn to go to the tips, but once my sack of coal–bits had been deposited at home I was out again into the fields for I could not forget the delight of the day on Spring Bank. I gazed around at the thickly wooded hills and realised that in our lives there had been beauty and simplicity. The resentment of my having to work down the pit began to lose its sting for I now realised how very fortunate I was to be living among these Staffordshire people. There was an extra warmth within me also for I knew that this new revelation, this process taking place within me had been instigated by association with a good man, an expert miner working at the coalface, a man who had shown me that inner peace was possible, even right down in the bowels of the earth. Abe Deakin had set my mind on a completely new course, a course which was gradually teaching me that my mind was capable of overcoming all difficulty. On this particular evening after my day on the dirt–tips picking coal, I seemed to be soaking my whole nature in the beauty that had been around me all through my growing–up years. I made a resolve to spend the whole of the next day on Spring Bank, 'alone'.

Yet again the sun came up with every prospect of a long warm day; I arrived at the summit quite early with a bottle of water and sandwiches, my pit clothes incongruous to the surrounding beauty. On the way up one of our Sunday School members reminded me that our Anniversary was only two Sundays away. Under my breath I muttered, 'There'll be very few new clothes for this Charity Day.'

In Silverdale, Anniversary was traditionally known as 'The Charity'. There was strong commitment and avowed dedication; sometimes obsession was mistaken for virtue, but, on the whole, Sunday Schools and Chapels were an influence for good and they produced good citizens. The fact that everyone went to work helped to establish an even mental climate, and to produce this blend in social and religious activity. As I lay on the hill with my face to the blue sky I reflected upon these things and found the memory of our village activities mingling with the charm and beauty around me. It had taken a strike to provide favourable conditions to make me aware of it all; we had looked at it as life had gone by, but we had seen little of it. We had never had time for reflection, we had never explored the beauty around us and no–one had been able or anxious to inspire us towards things aesthetic. How long I had been lying there I do not know but I was aroused from my reverie by a man's strong voice, 'Hello young man, enjoying the sun and fresh air then?'

I jumped to my feet in surprise, disappointed to a degree for I *did* want the day to myself. I knew the man's features for I had seen him around.

'Oh, sit yourself down, lad, I'll join you for a few minutes if I may?' I nodded my approval, but I was not keen to encourage the man to stay.

'My! it's a fine view, nothing to compare with it in the Midlands, neither is there cleaner air than that which blows over the top o' this hill. You are a Brown aren't you, one of the star piano pupils of Bob Herod? I know because your father pointed you out to me one evening on his bowling green when he was playing off an end with his brother, your Uncle Jim.' Now this added colour to the situation, my father and his prowess as a bowler. I began to show more interest and replied, 'Yes, he *is* a very keen bowler'.

'My name is Dan Roster and I live at Alsagers Bank. Fancy, four excellent bowling greens on Silverdale, great enthusiasm; I've watched your father play off some exciting games; have you noticed how even the women become excited when tournaments are played?'

This had stimulated my mind into action. 'Mr Roster, as I have been lying here since early morning my mind has been going over the features of our village and the surrounding mining communities and suddenly I have become aware of our beautiful environment and the interesting community activities; we have taken it all for granted. Are you a native then?'

'Yes lad, I've lived nowhere else. I'm glad you are noticing things around you and your village spares no effort to make life warm and interesting. This strike has cast a cloud over all our summertime activities. Do you remember the Morris Dancers, how they used to dance their way up through the village on summer Saturday afternoons, wearing their colourful regalia with ribbons and bells on their ankles?'

'Oh yes, I can just remember them. My brother George was one of the dancers. He was killed in the war. I remember Mr George Shelley, the blacksmith, being in charge of those dancers.'

Mr Dan Roster now became more enthusiastic. 'Harold, I've been trying to find out how long the bicycle parade has been in existence on Silverdale. I suppose the best–decorated bicycle competition gave the event its name instead of calling it a carnival. For you know it *is* a carnival. There must have been about eight tradesmen's carts all done up and presented as tableaux. Then came the clowns and those in fancy–dress, weaving in and out among the crowds, making sure that no–one escaped putting something into the box for the hospital. Hospital Saturday they called it, and rightly so, for it was a magnificent effort on everybody's part to raise money. It always reminded me of those religious festivals in far countries which I have read about in books; people forgetting their poverty for just one day in a year.'

'Yes, Mr Roster, our bicycle parade is indeed a festival. With all those fantastic disguises and face–masks and the complete abandon into which the whole village seems to throw itself, what else can it be but a festival? It has made an impression on me as I became older for I contrast the colour and excitement of that day with the poverty and drabness of our back streets and the struggle for existence which goes along with it for every other day of the year.'

Roster looked at me with serious expression. 'Carnival, Festival, Bicycle Parade, what's it matter? Silverdale folk collect a lot of money for the hospital and they call it Hospital Saturday.'

'I'm glad now that you came by, Mr Roster, for recently I've given a lot of thought to our communal activities. We accept Beech's Fair on Crown Bank as a normal thing, but I find that it is an exciting event in the pageant of our village life, just like Pat Collins and his collection of Fun of the Fair when Newcastle Wakes comes around.'

'Do you belong to a Friendly Society, Harold, have you ever been on one of their 'walks'? Now there's an interesting event if you like.'

'Yes, I am still a Juvenile Forester. My mother pays in a penny a week for me to keep me in some benefit or other, I really do not know all about it.'

'Harold, there seems to be as many Friendly Societies as there are chapels and churches.'

I offered one of my precious sandwiches. 'Go on, have one.'

'No thank you, but I'll accept a share of your water if you do not mind my mouth on your bottle.'

'Who ever heard of a miner objecting to that Mr Roster?'

I was now feeling pleased that this interesting man had crossed my path; it seemed that all he said had been complementary to my thoughts. Roster stood up as if to go, but he hesitated for a moment,

109

then said, 'Before I go, let me draw your attention to something else which I have watched growing in your village since I was a lad. You walk down any street during the evening in Silverdale, Knutton, Scot Hay, Alsagers Bank, Bignal End, Audley, or any other of our mining villages or hamlets, especially in winter, and what do you hear. You will not pass many houses before you hear a piano being played, someone practising singing exercises, others working hard at some brass instrument preparing for contest day. You in Silverdale have five Nonconformist chapels, two Anglican churches and one RC church. On winter evenings most of these places are lit up, practice going on for Sunday worship or for competition somewhere. Go and look at your day schools on winter evenings and they too are lit up for night school. Harold, what does all that mean? Not only great enthusiasm for music in various forms, but that there is a great reaching out for "culture". I think that there is a standard of culture in this district which is second to none. You mentioned the word "pageant" just now. Why, of course it is a pageant, a pageant of history passing right before our very eyes; why, think of the men from this district who have studied right deep into the night, gone to night school for years and at last obtained high certificates of qualification in the mines; they have obtained their First Class and Second Class papers. Then the great army of miners who are qualified firemen, deputies and overmen. It does me good to think of their effort.'

We sat in silence for half an hour during which my mind recalled the hours I had spent, standing outside one or other of our chapels listening to the choirs rehearsing 'The Messiah', 'Elijah', 'Israel in Egypt' and other of the well–known oratorios. I stood and took my last look over the plain and Roster asked if he could accompany me. 'I was on my way to Little Madeley, Harold, but I always come over this hill for a breath of this air. But if you do not mind, I'll walk down with you to your village for I have a call there; then I can return via Madeley.'

On our way down to Holly Wood, Roster took my arm and pointed in the direction of Keele village. It was an attractive view on that lovely day. 'Tell me Harold, did you ever see such a pleasing sight? Each time I walk across the ridge at Alsagers Bank I stop and look at the arc which embraces that hillside from Clayton right round to Keele, the rolling hills so well wooded, the Bluebell Wood and Davenport's Plantation and I wonder if there is a more satisfying sight

110

anywhere in the world. I assure you that I want to behold nothing better than that.' I did not reply for I knew how right he was, I too had felt exactly the same when I had stood on the north—west side of our village and admired the beauty.

Soon we were down in Silverdale and Roster again stopped and said, 'You have listened to me long enough, but I wish someone more educated than I am would write it all down, all that you and I have been talking about. Or, perhaps, make a film of it like those we see at the local cinema each week. Something which would remain as a permanent record so that future generations would know how we lived in this age and what we thought about the beauty around us.'

I thanked Mr Dan Roster and told him that I would convey his best wishes to my father. I had enjoyed his company and I too hoped that all the features which had constituted our conversation would not disappear too quickly, though I knew that the relentless forward movement of men and events would make all that we had discussed that day on the hill just a part of past history. Still, walking toward our house I embraced a hope that someone at some time would make a record of it all for the benefit of posterity.

Our mid—afternoon meal was already on the table and my father expressed his pleasure when he learned of my acquaintance with Mr Roster.

'Oh, so thou hast met Dan. Interesting chap, knows what he is talking about. He reads a lot, spends time in libraries and if anything is going on, thou canst bet that Dan Roster is there. I've been looking at thy notes and thy record of accidents and pit—disasters and industrial diseases. Where didst get all those items from? I borrowed a list of pit—disasters in this district for thee to look at when thou canst spare time from thy piano.' After the meal I noticed an old copy of the 'Staffordshire Evening Sentinel'. I learned that someone had lent it to my father; newspapers had been dropped from our weekly expenditure since the strike started. I retired to the front room to glance over this old copy of our local paper.

After forty minutes or so my father joined me and at once he looked into my face and exclaimed with some concern, 'Here what's the matter with thaiy, why art so glum?' 'I've just been reading this old copy of "The Sentinel" and even though we are all on strike we can still read of accidents and suffering in other industries.'

'Thou mustna let things like that spoil thy pleasure, for it seems to me that thou hadst a pleasant day with Dan up in th' fresh air. Here, here is that list I towd thee about. But dunna read it now for I'm going to sit here for half an hour to hear thee play. No studies and scales, play something with a tune in it; "William Tell overture", "Zampa" or "Tancredi". I may even sing a song or two to forget this strike.' I took out music and played for him something 'with a tune in it' and he sang a few of his old favourite songs.

After my father had left the room I finished reading the old newspaper and then turned to the list which he had acquired for me. Even before reading it, the sight of the full page of disasters sickened me. I had to wait for a few minutes before I could bring myself to study the complete list.

Pit disaster in the Potteries area	Men killed	Date
Bycars	5	January 29th 1859
Deep Pit	14	November 5th 1859
Brook House	5	March 2nd 1864
Clough Hall	5	March 1st 1865
Talke–O–Th–Hill	91	December 13th 1866
Homer Hill	12	November 11th 1867
Silverdale	19	July 7th 1870
Leycett	8	July 12th 1871
Berry Hill	6	March 12th 1872
Talke–O–Th–Hill	18	February 18th 1873
Bignall End	17	December 14th 1874
Bunkers Hill	43	April 30th 1874
Jammage	5	January 5th 1876
Silverdale	5	April 6th 1876
Apedale	23	March 23rd 1876
Leycett	62	January 21st 1880
Whitfield	25	January 7th 1881
Leycett	6	October 6th 1882
Apedale	9	June 20th 1883
Great Fenton	8	April 10th 1885
Apedale	10	April 2nd 1887
Mossfield	64	October 16th 1889
Diglake	77	January 14th 1895
Old Field	7	May 25th 1895
Minnie Pit	9	January 15th 1915
New Hem Heath	12	February 25th 1915
Minnie Pit	155	January 12th 1918

Slowly I read each item and as I passed from line to line my heart became more and more heavy. When I reached the last item on the list, 'Minnie Pit . . . 155 dead,' the whole incident came flooding back to me for it was the disaster which I remembered so well. 'The Minnie's gone up, The Minnie's gone up,' someone had shouted as they ran down the middle of the street that morning.

I could not take in the true significance of this document which I had let fall from my hand to the floor. Quite audibly I gasped, 'Oh, good God, what a list of dead miners, *what* a price for coal.' A cold shiver ran right through my body as I recaptured the horror of that Minnie Disaster and the dark pall of sorrow which hung over our whole area for many months. Halmerend, Alsagers Bank, Leycett, Knutton, Silverdale. Hardly a family was spared sorrow in some form or other through the loss of a loved one. My father had retreated for he knew what effect this list was going to have upon me.

And so it was that the mental buoyancy which I had experienced during the past few days became arrested and modified by the revelation of the tragic statistics of death. I now became conscious of the other side of the coin; the price of a ton of coal, the money paid by the consumer was now revealed to me only as a fictitious item in the estimates of our economy. The *real* price of coal was represented by the document which my father had given to me to read, the long list of the men killed and the sorrow of their families deprived of their bread–winner.

The miners waited week after week, month after month, trusting that their leaders would secure for them the best possible terms.

At long last the strike came to an end, but it was utter defeat for the men. Hunger and despair had brought defeat and humiliation, which in turn brought loss of self–respect. We had to return to the pit for lower wages and our shift was increased by half an hour. To make sure that no–one actually starved, the owners arranged a schedule of wages for each area; it was complex, but they all agreed to pay a subsistence allowance of six shillings per week to any adult miner whose wages did not reach two pounds per week. This allowance varied from one coalfield to another.

It was a humuliating condition which instituted a long, long period of inimical relations between pit–owners and men. Many called this subsistance allowance 'the most cruel blow of all'.

The strike had lasted ninety days and we were to start work on Monday. I arose early on the Saturday only to find my mother absent from the kitchen. During the night I had heard some disturbance and soon I learned that my mother was ill with quinsy. My father had already gone to the pit, and as my two elder sisters now worked away from home so the household chores fell upon my shoulders. I made sure that my mother was well-cared for and I cooked for my father and my two younger brothers. In accordance with the training we had all received in our house, I prepared all the vegetables for Sunday. For obvious reasons of safety my father had to be at the pit again on Sunday to help in the preparation for the resumption of work next day.

After breakfast as I began to prepare for cooking lunch, a knock came on the door. It was Mrs Framley. 'Oh, Mrs Framley, sit yourself down please.'

'Where's thy mother then?'

Pointing to the ceiling I replied, 'In bed with quinsy, go on up to her Mrs Framley.'

'Nay lad, I wunna go up in my clogs, th' tips 'll make a mark on thy mother's lino. I've only called to bring back a shilling which thy mother lent me last week. So thou't in charge of things then? Sunday dinner and all, how long has thy mother been ill?'

'Oh, only since yesterday. It started in the night.'

'What art cooking for meat? Dost understand it alright?'

'Oh yes, thank you. It's just a small piece of beef bought, I think, on the strength of pit starting again. But I haven't made a rice pudding for a long time so I am not quite sure.'

Mrs Framley watched me carefully and as I added sugar and milk she held my arm and said, 'That'll do for a pudding that size, now thy cinnamon and straight into thy oven. My word, lad, thou hast been well brock-in. Maggie Bloor [my mother's maiden name] knew how to break-in her brood. Look after her lad for she's a worker and thou art a good lad thysel'.'

Mrs Framley left, but that evening her daughter came and announced that her mother had sent her to care for us all till my mother was better. 'I'll come tomorrow morning to be with your mother immediately you and your father have gone to work. Don't worry about anything, I'll have your dinner for when you return from the pit, she said, reassuring me.

'How good of you, Elsie, and your mother too; thank you both very much.'

'Well, Harold', said Elsie Framley with her warm smile, 'you know Silverdale folks; if there's anyone ill or in trouble it isn't long before someone is heard coming up the entry offering help. People do tend to gossip sometimes, but you know it is the very same habit which gets the news around that someone in the street, or in the village, needs help, and response is never lacking. You know that don't you from the way your mother goes straight to anyone, once she knows that that person is in trouble.'

With this warming experience in good neighbourliness, I made sure that all my pit–clothes were ready for a fresh start at the pit after the long, long strike. Footsteps again sounded in our entry; it was Bernard Cliff. 'Well, Bernard, we are going back. The men have lost again. Something deep down keeps on urging me to try and get out of the pit, but, oh dear, what a hope. The miners have been on strike for ninety days and, thank God, it has been fine weather. They have been hungry and many have lost weight, so much so that their clothes hang on their bodies like rags on a peg. But they have had sunshine and fresh air which they would not have had down the pit. Whether or not it has been worth the price they have paid is another matter. There must be a balancing feature somewhere. I wonder how many of the men will find that, having been so long without proper food and nourishment, they are not strong enough now to perform their arduous tasks underground. At least the pit–ponies have enjoyed it in the Park Field and they have been properly fed, watered and cared for. No harm has come to the ponies Bernard for, you see, it costs money to replace a pony; but if a man is killed down the pit, or, if he is too ill or half–starved and cannot work, there is always another man available.'

My speech embarrassed Bernard I think, for he was anxious to change the subject. 'I wonder Harold, when can we two visit the top of the hill again?'

'You may ask that my old Sunday School friend. We did have a good day out on that day and the pleasure for me was so strong that after a day on the tips I made sure that the following day would be set aside for Spring Bank. Also I wanted to go and sit up there alone and think things out for myself. A man named Roster joined me and instead of thinking, we talked for hours.'

After Bernard had gone I sat quietly before the kitchen fire and the thoughts came to me, 'The enforced idleness has brought to me an endowment which I did not have before the strike. It has helped me to draw from the natural beauty which surrounds us some compensation for the ugliness and depression which goes with our work in the dark world underground. I am returning to Keele dip which is synonymous with danger and drudgery, but the beauty and fragrance of Keele and the surrounding country has brought delights and sensations which will now always be available; all this to a degree is compensation for the arduous life down the pit.'

My father entered the kitchen and disturbed my reverie, 'I'm afraid that thou must be prepared to go and work anywhere in th' pit tomorrow until things settle down a bit. There are plans afoot to give thee as much experience as possible. Thou art intending to go on with thy study, I suppose, so that one day thou wut take thy papers? There's an engineering course going to start at Stoke soon which would be useful for thee later on, think about it wut?'

7 Pumping

It was never intended that I should remain working at the face for more than a short period, just sufficient for me to obtain an insight into the kind of work involved. So it did not surprise me when my father told me that my first job in the pit after the strike could take me to any corner of any district. My first place of work was in Littlemine district where ponies were still employed although I was never given work which involved the pit–ponies. It was a pleasant experience for me to find such a flat road from the pit to the seam, hence the use of the ponies, but it was a different story once I was put to work near the face on a gradient which made it almost impossible for one to stand upright. My second morning in this district gave me my first experience with a timber hornet; no–one had told me of the existence of such creatures. Sitting waiting to start my ascent up to the steep dip with my lamp hanging near to my head, I suddenly felt this creature flying into my face. Frightened to death I jumped up and ran down the road outbye for a few yards for I had no idea what the creature was. The first men I met told me not to be frightened, but that I must not let it sting me in the wrong place or it might be dangerous. That night my father explained the hornets to me, 'They come down in the pit–timber, mostly in larch. They can be harmful if the sting is a deep one. The ponies are so sensitive to them that when they see one they stop, become agitated and if possible stamp on the hornet. Thou wutna see a lot of them lad; just keep out of their way.'

I was introduced to a pony, a piebald which I was told possessed very unusual intelligence. The pony's driver was another lad who belonged to my class in Sunday School. He said to me, 'Harold, once this pony stopped dead in his tracks; he wudna budge in spite of my strap on his backside. He even pushed back a yard or so; then, down it came, a fall of roof, just where we should have been with our three loads. I never gave him any strap again, in fact I brought him extra sugar from that day on.' I was fascinated with his story; I laughed and said, 'Yes, George, and we have the nerve to call them "dumb animals".'

Littlemine also brought me into contact with the pest of cockroaches; they were everywhere. With the dim light afforded by the Davy lamp it was not always possible to see when they were on one's body or clothes. These loathsome creatures occupied the bins used to store the horse–fodder. It was difficult to keep my snapping in a place where they could not penetrate. Again I was warned, 'Always examine thy snapping before taking a bite.' One day I forgot this advice; as I chewed on my first bite I thought of the cockroaches, opened my sandwich, and there it was, half a cockroach! I was sick on the spot and that was the end of snapping for me that day, the other half of the creature was in my mouth.

After Littlemine I was sent into South East so that I could become familiar with the endless rope system of haulage. My next move was into the pit–bottom with strict instructions from my father, 'Keep thy eyes and ears open, and I want thee to learn all the bell–signals on both Keele and South East engines. Pit–bottom work is as vital as work on th' face. Try and watch Bill Beeston, th' mon in charge of th' cage, it will do thee no harm to watch him for he knows his job. But dunna neglect thy work just gaping thou knowst.'

The pit–bottom was more congenial to me, for I could feel the cool air as it was thrown down the shaft by the powerful fans on the surface. Although the stink from the sump was always there, my work on the landing of Keele Dip and South East kept me well away from the shaft for most of the time. There was responsibility in each operation here for the pit–bottom is the central hub in coal traffic. I had to learn the clipping operation on South East landing and how to land the journeys from Keele; this required skill and accuracy with complete coordination with the engineman. And so I found myself in one new environment after another and gradually I was becoming a useful pitman; the pattern of my training was designed to stand me in good stead ready for the time when, later on, my responsibilities would be increased.

One Friday my father ordered me to go to work on Saturday night. The job was in Brown's Dip in South East district.

'But, tomorrow, that's Saturday and I'm on days this week.'

'Well, is that a complaint? Thou wut work thy day–shift tomorrow till one o'clock. That gives thee all afternoon and evening to get thy rest in and thou wut be paid time and a half.'

Time and a half meant nothing to me for I handed over all my wages to my mother who then gave me back my six shillings a week pocket–money. After a few minutes' thought I began to feel resentful that I should be expected to work down the pit two whole shifts out of a possible twenty–four hours. Again I was faced with the same question, 'If my father was not my boss at work, would it not be reasonable for me to refuse to perform this Saturday night duty after having only just come out of the pit at one o'clock on the same day?' My father was not anxious to tell me just what kind of work I should be going to on the following night.

After my Saturday morning's work I took my weekly bath after my meal. The zinc bath had to be carried in from its place on the outside wall of the kitchen. I went to bed to try and get sleep, but it did not come. I was not sleepy and in any case a lad was practising his cornet only three doors away. At nine–thirty that evening I was again on my way to the pit, but my father had called out as I left the house, 'Dunna forget thy two lamps, thou't need them.' I did not reply, but turned round and nodded to indicate that I had heard him. I kept on asking myself as I walked to the pit, 'Two lamps, two lamps, why two lamps?'

The fireman on duty was Jos Jones. He had worked on the opposite shift to my father at Number Six for many years. They not only understood each other, but they held each other in high regard as pitmen. He tested my lamps and chuckled. 'Hello Harold, so thy father's sent thee to do th' pumping.'

'Oh, so it's pumping then. Yes he has sent me and he says that I shall be working by myself?' I hoped that the tone of my voice would betray the fact that I was surprised and uneasy at the prospect of my working alone.

'Dunna worry thysel' lad, I shall visit thee.'

I waited in the pump–house at pit–bottom where Harry Wilson was reading his book. Jones arrived, and after a few words with Wilson during which time a glance was thrown in my direction, said, 'Come on, Harold, let's get thee to thy work.' Silently I followed the experienced miner through the pit–bottom and down into the South East dip. Soon we were at The Wall, an important junction in coal traffic from this district. Going down on one knee Jones said, quite characteristically, 'Eh–up Harold, let's have a couple of minutes bacca,' indicating a short rest.

'What is this wall here Mr Jones?'

119

'There must have been a road down on the other side and this wall bricked it off and at the same time gave support to this wide stretch of roof. On the other hand, there could have been a gob fire here long ago and the wall is the sealing–off wall. Thy faither and I had plenty of experience in that aspect of mining up at Number Six. We worked on opposite shifts in the same district for years. While I'm at it, I may as well tell thee that he often succeeded when some of us failed. He is a patient pitmon thou knowst.'

Soon we were at the top of Brown's Dip and I saw the water running in from the roof. Jones took me immediately to the large pool and let me try the hand–pump which was to be my constant companion for the whole night. Already I had taken on a bit of fright for most of the upright posts carried a thick white fungus which gave me the impression that they were two rows of tall silent ghosts. Now I was to be 'alone' in this place all night and now I could see the need for two lamps. Jos Jones took one of mine and fixed it on the jig–post, right in the centre of the dip and at once I realised that if he opened the door at the bottom of this dip and saw that light, he could assume that all was well with me. The official came close to me and said cheerfully, 'There's thy water, Harold, and all thou hast to do is to keep the level down.'

'So I have to keep this pump going till the pool is empty?'

Jones chuckled, 'Eh, lad, thaiy wutna empty that hole if thou pumpst till domesday. Just keep on till thy water looks a bit lower, then stop and have a minute or twos rest. Have thy snapping when thou feelst like it.'

I started pumping saying to Jones, 'Phew! This is going to be a nice way to spend my Saturday night off.' He smiled, but ignored my remark.

'You did say that you'd visit me?'

'Oh, ah, well, I'll visit thee before loosit.'

'Loosit, loosit, that's a long way off isn't it?' I felt that I wanted to prolong the conversation in order to keep him there as long as possible.

'Ah, but thaiy just keep pumping, but mind thou dustna get thyself in th' dark. Thou mustna bey in th' dark, working here alone.' With those words he left.

Pump, pump, pump; left–right, left–right. Dull and monotonous. My only companions were the sound of the water running down the side of Brown's Dip and the groaning complaint of the pump–handle. Suddenly I realised that I must have a marker to tell me how far the water had gone down; I slid myself down to the water–edge and placed a large lump of coal there. Struggling back I slipped, almost losing my light and, audibly, as if to an imaginary mate, I gasped, 'Thank God for that other lamp.' I was keenly aware of my precarious position, up here on the coalface, so far from any other human being. The goths became more and more frequent and more intense; they seemed to get nearer and nearer. [A goth is an explosion up in the roof. Generally they are innocuous and of varying intensity. They become dangerous only if they are quite near overhead and release loose lumps of rock from the roof.] Instead of ignoring them, I found myself stopping and listening hard. Since the day of the slip in Keele district, I listened to every roof noise I heard.

Suddenly there was a particularly loud goth and with it an ominous C–R–U–N–C–H of cracking timber. I suspected that there would be a fall of roof or another slip. Hardly daring to take a breath, I approached the post and saw that weight had come on and split the post from floor to roof. Quite loudly I cried out, 'Oh my God, how much longer to snapping–time?' My head became hot with fear and after an hour, an estimated hour, I left the pump as though I was making a major decision of my life and, addressing my imaginary companion, I said with a voice of authority, 'Well, I'm having my bloody snapping and to hell with the water.' Fear had really taken a grip of me and I felt desperate. I wondered why Jones had not yet visited me and my concern served to accentuate in my own mind how vital a visit was to anyone serving and suffering solitary confinement in prison.

My weary shift went on and the relief as Jones's light appeared cannot be described. I reported the post at once and, like the good pitman he was, he took chalk from his pocket, marked crosses on the post from top to bottom and immediately entered notes into his book, nothing left to chance. Jones observed my state of mind and asked sympathetically, 'Hast been frittened Harold?'

'I've heard enough goths to last me a couple of years.'

'It's twenty-to-five, Harold; thou canst knock off at a quarter to six.'

'But I anna got a watch, Mr Jones, I sold mine during the strike and I conna afford another one yet.'

121

'Here, lad, take my watch. I'm making for th' pit–bottom, I'll meet thee there'. This was a generous gesture and even in this mechanical device I felt that I had some sort of companion. The two rows of white ghosts did not frighten me on my way out for I was now walking to the pit, I was going 'outbye'. I was going to the shaft and then to the surface.

Back at the pump–house in pit–bottom I thanked Jos Jones for his watch. Then I blurted out to Harry Wilson the pumpman, 'I dunna want another shift like that one.'

Wilson turned with surprise, 'Harold, hast thaiy been up in Brown's dip all night alone?'

Before I could reply, Jones smiled and asked, 'Hast really been frightened Harold, hast really?'

'Yes, Mr Jones, I have been frightened and I dunna care who knows it. Each time I heard a goth, I thought it was another slip coming on.'

'Oh, of course, Harold,' said Jones seriously, 'thou wast up in that wasna?' He and I rode up on the cage together, and as we left the cage, Jones said, 'Yes, yes, thy faither should have thought of that slip.'

A new haulage system had been installed from the bottom of Keele dip to the far end thus eliminating the need for an army of waggoners. The two ropes of this new system, called Main and Tail Haulage, constituted a danger for they swung from floor to roof under great strain. I had worked for a short time in pit–bottom and again a short time in South East in order to learn different methods of work. Now I was helping at bottom of Keele Dip on this Main and Tail system, it was night–shift and I observed a light coming towards me. Here there came a young man about my own age holding up his hand, blood ran down his arm as he held it up in the light of my own lamp. He was in great distress and I learned that one of the ropes had come down on to his hand as he held the top of his waggon. Soon he was in the hands of a capable First Aider; his was one of several such accidents as the result of this new machinery. It was during this period that another of my schoolboy pals lost two and a half of his fingers in another part of the pit. It took a long time for me to recover from the shock of that accident. 'Poor Harry Ellis, what if *that* had happened to me; I could not then play my piano.' Such were my fears as I started each shift among the swaying, swinging ropes. It was a dangerous job I was doing.

'Watch thy ropes, watch thy ropes nagh. Never stride over them when they are in motion.' Such was the warning my father gave to me each night as I left home.

One morning, arriving home from night–shift, my father observed my state of mind which was betrayed by my countenance. He was preparing for a practice day with the Rescue Brigade; they performed exercises in putting on the rescue apparatus and spent some time in the gas chambers. He looked up to me and asked, 'Hello, who is it then, someone hurt?'

'No, not tonight, but I was thrown up into the roof by that blasted Main rope. It could have been nasty, but I had both hands free and I broke my fall.' I complained about the grim life underground, the harsh life in order that coal may be brought to the fire–grates.

'Thou't taking it too hard, lad. Remember this is pit–country and we are pit–folks. I've towd thee that before. Tragedy is part of our lives, like the dirt that settles on our windows from pits and potteries. We have to live with it, wash it off each day; then we have to retain what good things are left; the tragedy we must accept.'

Long timber and heavy steel girders were transported on low trams, which constituted a hazard when loaded and on the move. One night I encountered such a tram of girders. As I pushed it into position, the wheels passed over a bad joint in the rails; the girders slipped and trapped my left hand. At the hospital next day it was discovered that I had three fingers burst, one badly. Expressing my distress I pleaded with the doctor, 'Can I keep my fingers, please. I play the piano and I *must* have my fingers.' 'Oh, you'll keep your fingers, lad, but they are in a mess. You will have to take care and there will be a scar or two.' He was quite right for the scars are still there to be seen.

My Overman on day–shift was Mr Harvey Rogerson. Like Abe Deakin he was a man of quiet dignity and possessed an attractive manner in his speech. He was courteous in his dealings with the men and I deduced that he, like Jos Jones, had shared the more rigorous methods of getting coal; from this I assumed that he had worked in the Number Six Pit. I knew that my father held him in great respect and I was flattered as he stopped to talk to me from time to time, about my piano studies and often warned me of the uncertain future of the declining coal industry. He made repeated attempts to point out to me the wisdom of not having ALL my eggs in the same basket.

One day he said, 'Think about it, Harold lad; ask yourself, is pitwork worthwhile as a career for anyone who is really as keen, as studious and as young as you are?'

'I have thought about it, Mr Rogerson. There is already talk of another strike and that frightens me; I do not want to suffer that humiliation again. It is a dark future for us I think.'

The next day my father revealed that Rogerson had told him of his conversation with me.

'And did Mr Rogerson say that he had told me that pit–work as a career was hardly worthwhile?'

'Ah, he did Siree. All I can say to thee on that point is, what else can you do?' I returned to the front room to my piano practice. Soon I was joined by my mother who informed me that Rogerson had asked my father if he would consider having me trained as Keele dip Hooker–up, a responsible job at the bottom of the main dip. I turned round and asked sharply, 'And what did my father say to that?'

'Harvey wants thee as engineman later on when thou't old enough, but thy father will have nothing to do with it for it looks like favouring thee and thou knows how he feels about favouritism in this family.'

'Would it mean more wages?'

'No, same rate as thou art now having.'

The next day my father called me into the living room. I knew what to expect. 'Mr Rogerson asked me if I thought you could manage th' job of Hooker–up. I knew it was a challenge and I wouldna accept it at first. But he is thy Overman and he must decide. I told him that thou hast a grain or two of common gumption. Nagh, dunna tell anyone, just keep thy tongue between thy teeth. Thou knowst my standard and we conna help the way we are born and brought up. I have tried to pass on to thee what is expected of a man in this county and what kind of workmanship men in the village regard as honest. Perhaps thou wut pass it on to thy children one day, its nowt to be ashamed of. I'll have more to say to thee later about thy job. In the meantime dunna spread it around for that wunna do any good at all.'

My promotion to the job of Hooker–up at the bottom of Keele dip came two weeks later.

8 'Hooker—up'

The eventful Monday morning came and as I walked up to the pit I could already feel the weight of responsibility on my shoulders for I had spent the whole weekend thinking about my new duties and what the attitude of my mates at the bottom of Keele dip would be. I went down the pit early to make it possible for me to rush off down the dip immediately I saw the Road–doggy emerge from the cage. Already I had been told that if he should be absent, then it was my duty to clear the track of any obstacle before giving a signal to the engineman, 'Dip clear of men, coal drawing may now commence.' The Road–doggy soon appeared and I shot off to the bottom of the dip. The journey stood there all ready and the rope was at the bottom. I passed my hand over all couplings and made sure that the coal was not above the top of the waggons. 'Shall I be able to do this work as well as Redmond?' I asked myself over and over again. His example was to be my standard.

Within minutes the Road–doggy appeared. He looked straight at me and with his usual loud voice bellowed, 'Well, come on Harold, what art waiting for? Thou art th' Hooker–up thou knowst; I'm here, so the bloody dip is clear of men. Give the enginemon thy signal.' I leapt for my lamp and rang on the bell–wires, One—two—three. The rope slowly began to tighten and down I went on to one knee to examine each coupling as it went by under strain. I felt pleased that my first journey had been despatched safely. Ricket, the Road–doggy gazed at me and, with a tone of voice which I can still hear over many years, asked slowly, 'Hooker–up, eh now, was there a drag on that journey?'

Suddenly I realised that I had made a mistake. In my anxiety to make sure that all was well with the couplings I had forgotten whether or not I had checked that the huge iron fork, the drag, was suspended from the last load. Should the rope or any coupling break, I knew that loads would race backwards down the dip and possibly endanger life. No–one could see it in the dim light, but I was blushing with shame though I made as though I had not heard Ricket's question for he had spoken softly. Now with his usual fortissimo volume he shouted, 'Harold, was there a drag on that journey?' Suddenly, as though inspired by some friendly spirit, I realised that Ricket himself could not afford the risk of allowing a journey to travel up–dip with no drag on it. 'Er, yes, there *was* a drag on that journey alright. I checked all couplings before you arrived.' The second half of my statement was true, but the first half was a fib.

I never knew whether or not Ricket believed me, but never again did I allow the journey to pass me on its way up without a double–check of the drag, the huge iron fork which is suspended from the last load, designed to throw that load off the rails should the rope break. Fears raced through my mind that I should not measure up to the demands of this vital work, but as one journey after another thundered in, the short interval of time between my snatching out the shackle–pin from the empties with a shout 'UGH' to the order from me, 'Knock him up' became shorter and shorter. It went on non–stop till the bell rang . . . one—two—three—four—five. Snapping–time. I felt proud, for it was now my job to shout 'SNAPPING'. I heard it shouted up the shunt, then on to the first dip and so it rang on right through the whole coalface.

We all threw ourselves down on the ground, but I was too preoccupied to eat for a few moments. One of the waggoners came and sat at my side. 'What Harold, no snappin, artna hungry then? Has thy new job put thee off thy food? Thou must never let a job do that.' Johnny Wilson was still his cheerful self. He ate his snapping. Then, screwing up his newspaper covering, he turned and whispered, 'Thaiyt owraight kid, thou hast got th' speed. Take it from maiy, thou't make a good hooker and thou dostna let a loud voice worry thee. Good luck, Harold lad.' 'Thanks Johnny, I need a bit of luck for a day or two.'

Ricket was loud, he let us know that he was *really* in charge, but he knew what his job was and he exercised his authority just as efficiently as he accepted his responsibility. But already I suspected that he did not like my father. 'Well, young Brown, I didna think we should have gone this far without trouble. Thou art quick and thou dostna mind a bit of hard work. I think thou art settling in.' These remarks cheered me and by loosit on my first day I felt that I might one day receive the praise of Redmond D'Arcy himself. I set my sights on his pattern.

As my cage dropped through the shaft next morning I felt myself now to be a cog in the vast production machine called Silverdale Colliery. As I walked up under the lights of the pit–bottom, so a familiar voice rang out a friendly greeting, 'Hello Harold, my son, what are you up to these days?'

'Oh, hello Mr Deakin. I've just started hooking–up bottom of Keele.'

'Well done, well done. That is a responsible job. Now watch, Harold, Bill Beeston there, in charge of those two cages. Watch the way he commands those miners to get on to the cage, how he indicates that no–one else is to pass through the chain, and yet, he does not utter one word. He controls a vital part of this pit with economy of words and with no fuss. Now your job as Hooker–up is only one degree less responsible than his.' Deakin and I stood and listened to a group of night–shift men, with black faces, all in heated argument about the recent strike. It was all very high–spirited, but without rancour. If someone had pitched a note it is likely that they would have burst into singing together.

'Harold, did you hear what Clem Jones said? Did you get his point?'

'Yes I did. I also heard him speaking at a union meeting before the strike.'

Deakin's voice took on a more serious tone, there was urgency in what he said. 'What Jones said was correct. The men have lost only a battle, but it was an important one. They resent having responded so magnificently during the war to the country's appeal for more coal. Now they feel they are cast aside. Ah, more than that, they feel that they are being pushed into the gutter, crushed and humiliated. It will take more than one generation for the smart of this defeat to be forgotten and the *mark* it has made on the minds of these miners will *never* be removed. It will take many, many years for this insult to be erased. I shall not see it, but you or your children may. One day the effort of miners will be needed again; heaven knows what the response will be after this. How far in the distant future it will be I do not know, but mark my word Harold, there *will* come a day when miners and their skill will become recognised and their wages will be comparable with the danger and suffering they endure during a normal day's work. Now mark my words, lad, that strike has been a dark and ugly day in the history of mining.' The friendly argument in pit–bottom among the miners, then the serious approach to the topic

127

by Abe Deakin, led me to understand and appreciate the implications of the strike, things which had never crossed my mind.

I settled into my new work well and I came to take in my stride the hazards and responsibilities. It was in this job that I began to become aware of my approaching maturity. The opposite sex commanded some of my attention and it gave me pleasure as I realised that, should it become necessary for me to ask permission to 'walk out' with any young lady, my new job would be accepted as an indication that my prospects were fair. Rank in society and money in the bank counted for little when a young man faced his future father–in–law in the hope of being accepted. But let him say that he worked in the pit, was an engine–driver or worked on the face, or perhaps was in charge of a rip, then, he could be sure of acceptance. If the young man was a member of the Mines Rescue Brigade, then his stock would be high indeed. When character was being assessed in our area during that period in history, the vital question was, HOW HARD DO YOU WORK.

My work necessitated a rigid routine of one week on days, one week on nights. After several months when I was now known as a fairly useful pitman there came a shift on nights which was anything but normal. A fall of roof had taken all the men from dip–bottom inbye some three hundred yards to assist. Tom Burk, the Road–doggy in the main dip, was about to join them when our own bell rang 'Four' and a 'Five', indicating that the journey was off the road and help was needed. Burk stopped in his tracks as he entered the shunt to go inbye, turned back and came to my side. With his easy–going manner of speech he said nonchalantly, 'Harold, I towd thee didn't I, the Devil is working nights this week. Trouble inbye and now trouble in th' main dip. If I want thy help I'll ring.'

After only five minutes my bell rang again; one—two—three—four; one—two—three—four—five: 'Help wanted'. I grabbed my lamp and raced off up the main dip and I found Burk down on one knee in an attitude of contemplation. 'Harold, didst ever see owt like this; I'd like to tell that enginemon what I think of his speed.' We differed in opinion on this point, but it was such a complicated mess with timber knocked out and the roof unsafe, that it took us forty minutes to get the journey on the road again. Then Burk raised the question, 'Harold, who is going to hook–up? There's no–one there and there will be hell to pay if this journey stands still any longer after this long delay here. Thou hadst better jump in the last waggon.

Keep thy head down. If he stops before the bottom, that means no–one is there. Just use thy lamp on the bell–wires and knock him, then get thy head down again right sharp.' This I did and the journey ran in safely, but as I raised myself to get out of the waggon, a light was raised to my face . . . IT WAS THE NIGHT FIREMAN IN CHARGE OF THE DISTRICT.

Both Burk and I were duly reported. Realising the state of the roof where the derailment had taken place I ran inbye at once leaving Burk to cover my duties. I reported that timber was out, the roof still dropping dirt, and that the whole place was unsafe. Elias Green, the night Fireman, reprimanded me at once for not being at my post at the bottom of the main dip. But, as a good pitman, he thought about my report, returned to the place of the derailment and agreed that it required attention. Burk and I were subject to a lot of ribald leg–pulling for we were ordered to meet the Management on the following Thursday afternoon. Tom Burk pitied me for I had to deal with my father who was known for his rigid views of safety down the pit. But he took a fairly lenient view of it for he knew I had not received proper instructions before starting work in the pit, just like any other pitboy or miner. He now had to take me to task and warned me of other seemingly innocent activities in the pit which are strictly against the rules. He had relieved my anxiety by showing no anger at all. He heard my story then said warmly, 'Well dunna worry, lad, thou wast doing only what thou thoughts to be best.'

'Why aren't we given a list of the things we should and should not do in the pit?'

'You are quite right, lad, I've been fighting for years for a proper period of training both for pitboys and even for adults who come to swing on the rope for the first time.'

The dreadful day came when we had to face the Under–manager in his office. Elias Green who had reported us, Harvey Rogerson who had recommended me for the responsible job and my father who, as Overman, had been invited to attend. My own opinion was that my father had asked for permission to attend to make sure that I was not favoured in any way. The Under–manager looked straight at my father, 'George, your Harold is in trouble. He *did* ride down a main haulage road in a waggon, strictly against all rules. Harold says that he did not know that it was against the rules and I believe him. Elias Green reported th' lad because, like thee, he strives for safety underground. Shall we just fine these two men and let them go with a warning? What do you think George?'

'That must be your decision Mr Burgess. You know my views about the weakness of our system under which a lad of fourteen goes into the pit without proper training. I am not here officially and am no part of this enquiry. If our Harold has broken the rules, and there is no doubt about that, he must be treated like any other lad. Don't favour him; that's th' last thing I want.'

There was a pause. Mr Tom Burgess seemed to want all three men to express an opinion. Not only was he a very fair man, he was also astute. At length I was given the opportunity to explain why Burk and I had decided that it was very necessary for me to ride down. I obeyed my father's instruction not to say too much and I ended my recital with the words, 'We both did only what we thought best in the interest of the pit.' My father was again invited to express his opinion, though he said less than I did, but made it clear that at some time or other everybody makes mistakes. He also drew attention to the fact that I did my best to get the journey away from the dip—bottom and re—establish coal—drawing.

Up to this point Harvey Rogerson had remained silent, but from time to time he gave me a sympathetic glance. Mr Elias Green had adopted correct procedure, but no—one was concerned about Burk's effort and mine to get coal moving again. Memory does not serve me well enough for me to set down every detail of that meeting, but what Mr Rogerson said has remained fixed in my mind throughout life. Burgess nodded his permission as Rogerson indicated his wish. 'Mr Green, I am present today because I made a request to hear the details of this incident. It was natural for me to take an interest since I was responsible for this lad being given this exacting work. Would you say that he is good at his work Mr Green?'

'Oh yes, I've nothing against him personally. Harold is a worker.'

'Would you agree that we men in charge sometimes commit errors; we are sometimes obliged to depart from the printed word of the book, especially when under pressure? After all, you were under pressure on the night in question.' I marvelled at the technique Harvey Rogerson was employing to induce Mr Green to agree to all he said. Now, with stronger accent and more tenseness he said, 'Gentlemen, I put Harold Brown down at bottom of Keele dip. I knew that he would make mistakes as we all do. It has been pointed out that no–one is perfect and this lad made a serious mistake; he has broken a pit–rule, but it was not neglect and the lad did not know the extent of his offence. How many of us can boast of being immune from neglect in our pit–life. I am defending this lad for his journeys have been turning round at dip–bottom a few seconds quicker than average. Those of us who have to answer to "average" output know what this means. Output is the dominant feature of our work, it is drilled into us by our superiors. This lad is a good worker, let us not break his spirit not even by fining him. Give him a caution and let him go; encourage him and let him see that he is working for good masters.' I felt proud of this praise when I was there to receive punishment for my misdemeanour.

Burgess took in a deep breath, looked at all three officials in turn and said, 'You will agree that I have a responsibility. Harold will be fined a pound; Burk must pay ten shillings for watching the lad get into the waggon.' He had referred to me as a lad. Throughout the whole proceedings it had been stressed that I was a lad, but down in the bowels of the earth I was expected to do a man's job and behave like a man. My wages were those of a pit–lad, less than two pounds a week. Returning home I went straight to the rocking–chair in the front room. There I was able to close my eyes and consider all that had taken place in that office. I had begun work in the pit at fourteen years of age at the wage of thirty–five shillings a week. I was now approaching seventeen and regarded as a useful pitman and my wages were not yet two pounds a week because of the gradual decline in wages. A fine of one pound meant that I should lose over half of my wage. But it was not this which occupied my mind in the arm–chair. It was something which had given me great pleasure, for I had observed carefully how each of the four officials employed tactics which would give a plausible appearance to what they had to say. As I watched every move and listened to every word I seemed to be captivated by their respect for each other, no word of insult nor criticism was uttered, each recognised the status of the others; dignity prevailed throughout. It would have been incongruous for any miner to disparage in the slightest degree, the character, skill or conduct of another miner. They were all in it together and they behaved according to the high standard which is always maintained under-ground. A gratifying warmth flowed through my whole being as I realised that it had not, after all, been such an unfortunate affair.

After our meal my father related the fact that Burgess had told the officials, after Burk and I left, that he would adopt the same strict attitude to offenders against safety–rules whether it be boys, youths, minor officials or senior officials. I could see that this display pleased my father who went on to say that Harvey Rogerson said that he would give me some overtime to compensate for my loss, but I persuaded him not to for obvious reasons. My father's pleasure lay in the fact that my mistake had brought into the open the text of his own constant pleading and preaching for organised training before actual work down the mine; it had accentuated all he had said on the matter. The means by which this revelation had been made did not matter, even though his own son had been the instrument; I knew that my father was pleased.

The meal was over, but there was still a cup of tea to come. We were back in regular routine now and I never took that for granted once the strike had ended. As my mother poured the tea she interrupted my father's conversation, 'Yes, a good worker, doing well at his job, trying, trying, trying all the time and they damned well fine him, more than half of his net weeks wages.'

My father ignored this outburst and said quietly to me, 'Nagh Harold, forget this incident; it has served a good purpose, but thou must ask me or someone else if thou art ever unsure of anything at that pit. Ignore the tantalising leg–pulling of the men, they dunna mean thee any harm.'

'Oh, I know that, they have all been so warmhearted in their gibes at Burk and me, but I've not neard one remark which echoed "Serves them right". Everybody has been decent about it.'

My father turned to my mother, 'Maggie, it *is* hard and Harold must pay the fine himself; it must come out of his own pocket– money. I will provide the pound, but he will repay me at a shilling a week. He must be able to tell folks honestly that he himself is paying for his own mistake, the whole affair has been good for him. The smart wunna do him any harm and the experience will drill it into his mind the importance of "safety". Now let that be the end of it.' But my father was not aware that he referred to one lesson, while I was warmly embracing the quality of another lesson I had learned that day in that office. Dignified conduct among a group of senior officials at a coal–pit, all highly qualified by experience, but holding forth with a difference of opinion on a technical point. Their utter respect for each other claimed my deep admiration for every one of them, including Elias Green.

Bernard Cliff and I walked around Keele that night during which time we both expressed a wish that the conditions which had given us such happiness in childhood could return. We were both still young enough to cherish the memory of days when family, home, community activities, Sunday School associations and the dedication of adults who showed concern for us, all combined to give us a strong sense of warmth and security.

Burk was absent from work that night. The reason was entirely his own business, but he was there the following night with his infectious smile and good humour. We both accepted the jovial banter from every man in the pit on my very expensive joy–ride down Keele dip and Burk's ten shilling–fee for the pleasure of sight–seeing, watching a pitboy get into a waggon for a ride. It was a relief to participate in this good–natured exchange, but I ignored any remark which cast reflection on Elias Green and his report, for he had displayed admirable qualities and HE WAS A GOOD PITMAN. The loss that I had suffered was much, much more than a pound out of my wages. It seemed that I had put a mark of disrepute against my father's long record as a first–class miner. The generous attitude he had was displayed in refusing to become angry over the report, his fairmindedness in making me responsible for the price of my mistake and his estimate of the final result. The unpleasant nausea which I experienced during the first twenty four hours following my ride in the waggon was now gone. What had started as an irritating episode had ended with an increased admiration for my father and a much deeper respect for those in authority at the pit.

The months followed and, like anyone else in the pit, my usefulness increased with practice and experience. Mr Harvey Rogerson continued to warn me of the bleak outlook throughout our coalfields, but he still encouraged me in so many matters. My studies continued and the responsibility contained in my job increased my sense of diligence; an eye always on the rails with my lamp close to the ground, my ear constantly on roof–noises, listening for any evidence of 'bitting' which was a warning of imminent fall of roof. My father now gave me regular talks on pit technique and encouraged me to ask questions.

Some months after I had been fined he called out to me as he arrived home from work, 'You will be working on Saturday night.'

'It inna pumping again, I hope; where is th' job then?'

'Never thaiy mind and dunna ask questions, thou wutna have to kill thyself with work on that shift.'

'That will be a change for me to have an easy shift,' I replied saucily.

When I arrived at the lamphouse on the said Saturday night the man in charge silently handed me *two* lamps, both of the type used by officials, lamps fitted with a screw–device at the base for the turning down of the flame when testing for the presence of gas. This mystified me as I rode down the shaft with only one other passenger, my father. I knew better than to make enquiries. After a moment in the pumphouse at pit–bottom during which my father whispered a few words to the pumpman on duty that we would be working in Spencroft heading. 'A heading' I mused. 'That is a new road being driven in, what *is* the job?'

The pumps were silent, all the pit was silent, there could have been only a dozen men in the whole complex of the pit that night. As I passed under Keele dip engine and glanced at the silent South East engine I became aware of the awesome silence everywhere and I thought of the thousands of miles of pit–roads running under the ground throughout the whole country. My imagination was at work. 'What a world this is, all these miles of narrow passages under the crust of the earth. Men working like ants, no–one on the surface ever sees or hears it.'

The reason for *two* lamps and the special type to boot, still occupied my mind. 'What are these two lamps for, we have three lamps between us, all special ones. Why is that, where are we working then?'

'Just keep thy eyes open and thy wits about thee and perhaps thou't learn.'

Spencroft district was a small area, recently developed and only a few hundred yards from the shaft. We suddenly entered a very steep gradient. My father halted only half way up this steep dip and rested on one knee. 'Nagh, get into th' habit of stopping half–way up any steer like this one. Thy heart will last thee longer if thou doest that. Listen to me panting like an old horse. Take good care of thy heart lad.' Once at the top, the going became easy and again my father stopped, pointed to a pile of tools, all neatly stacked, but uncovered.

'Hand me that small shovel and thee bring one of the picks. Now, here is an important lesson for thee, an object lesson I should say, a lesson about pitmen. These tools are not mine and I have no right even to touch them. It is an unwritten law in the pit, YOU DO NOT TOUCH ANOTHER MAN'S TOOLS UNDER ANY CIRCUMSTANCES WHATSOEVER. This law is as sacred as anything thou't find in thy Bible. A man's tools are his means of earning a living, so, leave them ALONE. Unless of course he gives thee permission. Now remember that as thou doest remember thy birthday. Dunna ever, ever touch anybody's tools. Knowing that I should need them I asked th' Butty of this set of men; few miners will refuse thee if thou doest ask. He wunna lose anything for obliging us with his tools.'

Within minutes we were at the heading. There it was, a blank wall of dirt. It was not hard rock, just compressed dirt. I marvelled at the care with which my father tested every square foot for gas. Now I could see the reason for the special lamps, but my curiosity became too much for me. 'What are we here for, what is it we are going to do?'

'We are going to thirl, does that satisfy thee?'

'No, what is thirling?'

'It is an important operation. You have learned that there is usually a road in and a road out when men are at work in the pit. Here there is only one road, hence its name, "heading". There is another road on the other side of this wall. It is the same system as that used when two teams of men work towards each other when making a tunnel through a mountain, so, here I should meet, some time tonight, the road on the other side of this wall, that is of course, if the surveyors have done their work correctly.'

'Why is it important?'

'Thou art asking too many questions and there inna time for me to go on talking. I dunna know exactly how far I have to go, what conditions I might find, and there may possibly be gas to deal with.'

The operation was tedious with nothing for me to do, but to keep my light trained on to the point where my father gently picked away the dirt, only a bit at a time. This I did for most of the time, but when the dirt accumulated at his feet I took the shovel and threw the dirt behind us, well out of the way. My patience began to feel the strain and through my mind ran the thoughts, 'If only there was a bit of pumping, a waggon or two to push up, or, if only I could hear the sound of bell–signals in the distance.' But no, I had to stick it out.

135

Once, I just had to let go and I burst into the singing of one of my favourite out–of–doors Anniversary hymns. 'Come ye that love the Lord' to the tune Hobbs. My father stopped, put down his pick and looked at me straight in the eye.

'Here, shut up wut. What art thinking on? I towd thee to be quiet.'

Then, perhaps realising the strain it was for me to stand there in deadly silence, just holding a light, his tone changed, he dropped down on one knee, 'I know that it is a strain for thee; it is bound to be a trial, but I am having to listen to every sound which may come from the other side. Just keep thy light steady and be very quiet, lad.' This softness in his voice and his understanding of my boredom changed my attitude towards the whole operation.

I began carefully watching his face and observing his concentration as he picked gently at the wall of dirt, then tested for gas, then placed his ear closely to the wall and listened. Although completely ignorant of the nature of the night's work, I began to appreciate that he was doing something of an urgent nature and mentally I began to participate in the operation. Questions arose in my mind as to what thirling really was in principle; there was an element which I felt my father was keeping from me. I was still mystified. 'What *is* at the back of this thirling; was my father invited or ordered to do this work? There are several other men at this pit just as experienced and able as my father, why must *he* be doing it. My father has recently been promoted to Overman in South East, Spencroft is not his district, so why must it be him working quietly through Saturday night? He is on standing wage and working Saturday night will not bring him another penny in wages. Why of all men in this pit was my father sent to do this job?' Questions, questions, rushing through my mind.

After three hours of this slow monotony, we sat and ate our snapping. Snapping–time on an ordinary shift is quite an occasion, twenty minutes break after hard toil and the satisfaction of food and the feel of cold water running down one's throat; all this is a pleasant experience because of the utter contrast. But now, a chance to sit down and enjoy sandwiches after three hours of absolute stillness in a dark, remote corner of a pit, this was luxury indeed. But still the questions loomed up in my mind. 'Father, why do we need *three* lamps?'

'Goodness, lad, is there no end to thy questions? When men are working alone in a quiet part of the pit, the extra lamp inna amiss and we did not have far to carry it from the bank did we. It is just an extra precaution in case.'

I took particular note of those two words 'in case' and I wondered, *'in case of what'?* 'Father, why did you place that third lamp further back and just round that corner? There's no benefit to us with the light around the corner and that far away.'

'More questions is it then. Why doest think lad? Should anything happen in this small confined space and our lamps are knocked over and we find ourselves in the dark, that other lamp would be useful to get us back to th' pit–bottom. Nagh come on, let's get th' job done.' We resumed the slow monotonous procedure.

Suddenly my father stopped, he dropped the pick, remained motionless, then said in a low voice, 'Nagh, not a sound, I think we're through. Thou hadst better get down low; go back a bit in case. I will hold th' lamp in my teeth and use th' pick with one hand.' He needed his left hand to support himself. He now held the pick with his right hand, holding it near the blade. My heart began pounding for again he had said, 'just in case'.

Obeying his command I retreated slowly and lowered myself on to one knee. This order 'go back a bit' made it obvious that he anticipated some danger, but in spite of this possibility it meant that there was now action and I jumped at the opportunity to be part of it. From that moment my feelings were of nothing but absolute pride as I watched my father's care and concentration with his lamp gripped firmly between his teeth, a common practice in mining. A surgical operation could not have been performed with greater exactness. My nerves tensed, I asked myself the question, 'What *is* going to happen?' Then came the sudden high–pitched whistle as the air rushed through the tiny hole. I leapt forward and held up my light.

'Well done, lad, now keep thyself down a bit yet.' He held both hands now over the tiny hole, allowing it to become larger only by small degrees. Gradually it became as large as a dinner plate, then large enough for a man to crawn through. My father stood away, gazed at the hole and said with an appearance of satisfaction, 'I think we've won, my son, ventilation seems to be normal.' As if to grant me recognition of my patience he had said *'we've* won'. Then, as an extra bonus he uttered words I had not heard from him before—'my son'. Abe Deakin had used the expression, but coming from my father it touched the sentimental point in my nature and I enjoyed the luxury to which I was unaccustomed. I felt that my patience had had demands made upon it, but now I was sharing the gratification. My pride in my father was such that I suddenly realised that all my complaints and resentment at having to work underground for my living were dissolved. Before we left he stopped and took a final look at the hole; I remained silent for I sensed that it was a moment he wanted to himself. To this day I have never been able to learn exactly what possible danger he had in his mind.

'Now, see that those tools are replaced exactly as you found them.' That was his last command and soon we were at the pit ready to ascend the shaft. 'No breakfast for me, tell thy mother. Just a cup of tea.' I took the three lamps, handed them in and wandered slowly homewards. I sent my mother off back to bed, cooked my own breakfast and stayed to make my father's tea, but my mind was back at the heading. I was glad to be alone with my thoughts. Within me I was warm and relaxed for I realised that another compensation for the darkness and dinginess of the pit had been afforded me; drinking my tea, I struggled to place in proper perspective the quality of the experience I had shared. I tried to find suitable language to describe my inner feelings, but words did not provide the medium of expression for such rare moments. Sitting in silence I mused, 'What a man my father is, quiet application at work with no fuss. How real it all is down there, every action worthwhile and of significance; no veneer, self—boasting or ostentation, no room for pretence; it is truly a man's world.' My pillow was particularly soft that early Sunday morning for my whole being was at ease with a sense of intense gratification and fulfilment.

9 A gob fire!

Although the long strike of 1921 was now in the past, I could still hear grievous complaints, groans of resentment and bitter dismay at the appalling conditions of settlement which had been forced upon the miners by the pit–owners. Harsh words were used; some said that the settlement was barbaric, others said that it was serfdom come back again. 'A fraud, a sham, utterly dishonest' was a typical expression used when men gave vent to their feelings and exchanged personal views. Waiting for the cage, in the queue at the lamphouse, or during snapping–time, the complaints could be heard each day. The men were suffering more than mere defeat, for their standard of living had been drastically reduced. One well–known choirman once said with serious voice, 'They oppressed them with burdens and made them serve with rigour.' Thus he reflected the sentiment expressed in Handel's great oratorio, 'Israel in Egypt', the story of the slaving Israelites in bondage in Egypt. My feelings went out to these miners; it was not surprising that one heard rumours of another strike and each time I heard such a suggestion, the familiar cold shiver passed down my spine. I had no doubt that it was a psychological reaction, an echo of mind from experiences in early childhood during strikes. So trenchant were the marks made on my mind in early years.

One day Harvey Rogerson came to me and repeated his warning of bad times to come, but always his remarks carried some balancing feature. 'Harold, lad, the future looks bleak and pit–work is ghastly and humiliating, but it gives a man an opportunity to prove his manliness.' The grim outlook did not give me zest, nor increase my enthusiasm for my work and often I reflected upon the cavalcade of accidents and misfortune which had fallen within the orbit of my own experience since I entered the pit. Continuously my attention was drawn to the misery and dejection on the faces of the men; those who normally possessed high spirits and good humour could not now raise a faint smile. Apprehension existed everywhere. My father, on the other hand, evinced his pride in my work and, for some reason I could not understand, he made a point of making complimentary remarks about my performance. But even this did not erase the gloom from my mind. The possibility of getting out of the pit again occupied my thoughts from time to time.

One day my father beckoned me to the front room. Cheerfully he said, 'Mr Rogerson wants thee for enginemon later on when thou reaches seventeen. He says that it will be better for thy piano fingers, Siree, but I anna having anything to do with it. Men will say that I'm favouring thee. If Harvey wants thee, then it is up to him. He is thy Overman in Keele. Thaiy costna do that work till thou art seventeen. Dunna let it go to thy head; anything can happen in the meantime, and keep thy tongue between thy teeth. Thou knowst what gossip can do. Just wait till thou art towd officially.'

Some weeks later a note was clipped to my lamp, and this brought me to an interview with Mr Rogerson. Courteously he offered me a seat and said, 'Harold, you have done your work well as Hooker–up; I knew you would succeed. Now, your new work will involve a very powerful engine and that power will be in your hands. I ask you seriously never to abuse it. The success of our effort to increase production relies upon the efficiency of the main dip engine and its drivers. Never, never risk getting your journey off the road or doing anything at all which may endanger your workmates, merely for the sake of a bit of extra speed. Efficiency in work, a quick turn–round at dip–bottom and an ever–alert ear to the noise of your engine; you understand most of that already. The rest is commonsense and I know that you have your share of that. Good luck, Harold, when the time comes, but I shall speak to you again before you begin your training. I shall be behind you in difficulty, but I shall expect you to be responsible and to answer for any trouble which arises from any foolhardiness. I know that you will do your best.' Swelling with pride and satisfaction, I thanked my Overman as best as I knew how to.

It was difficult for me to take it all in. Someone, a personal tutor, was going to teach me a particular skill. A new skill in pitwork was going to be taught to me by a real teacher. Also, that teacher was going to be Fred Askey, one of the pit electricians, a man for whom I had great respect and admiration. He was a quiet man with a serious turn of mind and I liked his company. He was a member of a well–known Keele family, and he always impressed me with his talk of country–life and activities. I always imagined that he would have made a very efficient farmer. Now I could hardly wait for the day that

I should put away my pit–clogs. Boots were a necessity for accurate footwork on the brake. I anticipated the clean enginehouse, coolness at pit–bottom where the engine was housed, electric light around me and no need for the glimmer of the Davy lamp. Straight off the cage to my work and up the shaft on the first or second rope at Loosit; this was going to be a privilege, even luxury.

Not long after my interview with Harvey Rogerson a normal day was proceeding at bottom of Keele dip when there occurred a very heavy fall of roof inbye, half way between main dip bottom and far end. It was sufficiently serious to bring the Under–manager down. The telephone rang and Harry Wilson, the pit–bottom pumpman, said to me, 'That you, Harold?'

'Yes, Harry, what is it?'

'Has Mr Burgess reached the fall yet?'

'Yes, he went in just now.'

'Well, will you go in yourself and make sure he gets this message?'

'Yes, I will go. What is it, Harry?'

'Now make *sure* that you deliver the message yourself. Say that the place in South East has turned sour and will he go straight to it, right away.'

'Alright, Harry, I'll make sure he gets the message, rely upon it.'

There was no journey standing in dip–bottom, every man had gone inbye to assist at the fall. I took down my lamp and ran in to find a mountain of dirt and stone down and immediately uttered under my breath, 'Thank God the men were not walking out when that lot came down.'

Burgess was at the top of the huge pile examining the great cavity in the roof with his electric light. He slid down looking very concerned. 'Mr Burgess . . .' Before I could get my message out he retorted. 'Harold, what art thou doing here, dunna tell me thou hast got trouble.'

I beckoned him aside and said softly, 'Harry Wilson rang. He says that the place in South East has turned sour and would you go to it right away.'

'Oh my God, wait a minute Harold, I'll come out with thee, but first I must knock these men off.' He walked over to the Fireman in charge of the district and spoke in a low voice giving instruction for everyone to be knocked off. Naturally I assumed that the fall was so big that it had put an end to coal being drawn for the rest of the shift.

'Harold, thou art young, thou hadst better get up to the face, tell everyone to knock off, then make thy way up to th' pit. Wut volunteer to do that?'

Before I had time to reply a familiar voice spoke up; I had not been able to identify the men standing around waiting for the instructions from Burgess. This was the voice of my friend Abe Deakin. 'I'll go with Harold,' said Deakin, 'there must be two.' 'Oh thanks, Abe, thou knowst what to do, there's trouble in South East.'

I had not taken in all that Burgess had discussed with the Fireman, but I had heard the words *gob fire* and suddenly as Deakin and I rushed along the coalface it came into my mind that everyone had been knocked off because of the gob fire and not because of the fall. Fear took a grip of me and this was made worse when Deakin said, as we reached the end of the coalface, 'Well go on, Harold, off you go up that dip to the pit. You will get on much quicker than I can.'

The words, 'Gob fire, gob fire, gob fire' were throbbing through my brain. The dip appeared to be twice its steepness now. As I reached The Steer my panting was now more of a painful groan as I forced my legs forward and upwards. Lights appeared ahead of me and I assumed that they had joined the main dip at plate–landing, men working in the air–road. As I came near to these men the last one became aware of my hurrying; it would have been quite impossible for me to push past them; also I could see that they were older men by their slow pace. The sound of my loud grunts as I struggled for breath to sustain my effort brought attention from the last man in the group. He shouted loudly, 'Eh, just stand aside a jiffy, there's someone here, a youngster running up. Come on past, young 'un, thou't quicker than us. Come on lad, get thysel' up to that pit.' THAT MOMENT I SHALL NEVER FORGET.

I could not afford the breath nor the effort as my lungs laboured to keep pace with the demands of my struggling body, but I made myself look aside at him as I passed him with bent body. With grunting voice I ejaculated, 'Thank you mister.' On and on I raced till the welcome lights of the pit–bottom greeted me. There was a queue for the cage. It was not loosit so I only could assume that others had been knocked off as well as Keele district. I was amazed at the almost complete silence; no ribaldry which usually helps to lift the boredom of this unwelcome waiting for the cage; there was just a sombre silence as we shuffled forward and the queue in front became shorter. Each time the cage disappeared I counted the men in front of me and at last I felt the touch of the rough hand as, Bill Beeston, the man in charge of the cage, let me through. Not minding who heard it I sighed deeply and said loudly, 'Thank God for that, shall I be glad to get my feet on to that cage.'

From the moment I had heard the word gob fire and realised the significance of Burgess being fetched away from the fall of roof, my whole being had been in the grip of fear. The stink from the sump had always nauseated me, but today, the obnoxious smells at pit–bottom were very welcome indeed. The next cage down was MINE and I cared for little else at that moment. The tension there was in that pit–bottom cannot be described, but I recollect very clearly thinking to myself as my eyes were glued to the shaft, waiting for my cage to appear, 'What if anything were to happen right now, just as I am about to get on the cage and be taken to the surface and to safety . . .' Such were my fearful thoughts. In spite of this tense atmosphere I observed that there was no pushing, no–one tried to jump the queue; the display of discipline matched the situation. But here was my cage, at last. The first man to alight from the cage was, 'my father'. Being Night Overman in South East he would have been summoned from his bed on first news of the emergency in his district.

I was near the back of the group of men about to enter the cage and I saw that my father had spotted me. He stood aside as he stepped off the cage to allow the rest of those men who had occupied his cage to pass through. He gazed at me questioningly for a second or two, as if not sure what to do; then he touched my sleeve. When I did not respond to his touch, he laid his hand firmly on my shoulder and said, 'Here.' He beckoned with his head that he did not wish me to get on the cage. But I still made as though I was sticking to my place among the men moving to the cage. Beeston was more alert to the situation than I was for he had already signalled with his hand for another miner to take my place. I stood aside for a moment and appealed to my father with my eyes, but he just stood there and waited for me to join him. 'But this is my rop,' I blurted out with some emotion. I badly wanted to disregard his indication that he wanted me to go to him. I could not believe that he would wish me to remain down the pit one moment longer than was necessary.

'Here, Siree, I want thee, dunna get on that cage.'

I knew better than to disobey and his order was direct enough for

143

anyone to understand. So, with downcast eyes and troubled spirit, I turned aside and watched *my* cage rise up into the shaft. My heart sank, but I have never forgotten the look of sympathy which Bill Beeston gave to me as he noted my frustration. Beeston was a big, burly man, quite often brusque, but now I could see that he was very human under his large frame and lumberjack appearance.

My father entered the pumphouse without a word and I followed him. After a moment of silence I almost shouted, 'Why can't I go up the pit, everybody else is going up?'

'Everybody else inna going up, I anna going up for one, I'm just coming on. Nagh, listen, I dunna want any fuss. Listen to me. Get thyself off wom, have thy dinner while thy mother cuts thee some snappin'. Bring plenty of water because I didna stop for any. Come back to th' pit and join me in South East. Bring two lamps and wait at The Wall till I send for thee. Nagh, dost understand me?'

'Yes, have my dinner, bring some snapping, plenty of water, two lamps, wait at The Wall.'

'Well, get thysel' up that pit now.'

'Have I *got* to come down again with that fire in South East?'

'Thou just do as thou art towd. Thou dost understand that dostna?'

It was not my habit to swear; I had used the odd swear–word when trying to impress my workmates or to boost my ego, but it was only on rare occasions. This situation was too much for me, I walked slowly from the pumphouse and quite aloud said, 'Bugger it, blast the bloody gob fire.'

I walked right past the queue to the cage; I heard a murmur from some young man to the effect that I was jumping the queue. I was feeling far from pleased and this snarl was like a light to gunpowder, for I had already waited my turn for the cage. I turned to the young man, took a hold of his shirt with both hands and pulled him toward me, I shouted into his face, 'Who art thee then? I dunna know thee, but if there's something thou wants to settle, we shall both be on th' bonk in five minutes.'

Beeston with his ever–alert eye saw the trouble and in a tone which proved his undoubted authority, shouted, 'Here, young Brown, come up here, thou hast waited once for thy turn.'

The young man who had offended me saw the mistake he had made and he rose to the occasion. I never knew whether anyone had told him of the circumstances, but he shouted to me as I waited with the group to go up the shaft, 'Sorry, Browny lad, I didna know, sorry kid.' That was enough for me, I turned, put my hand up to him and gave him an apology of a smile; it was all I could muster under the circumstances.

On the cage I turned to a man I knew as we rode through the shaft and said, 'Fancy that, I've got to come down this bloody pit again.'

'Ah, I thought as much, Harold, when I saw thy father talking to thee.'

Stepping from the cage he stopped, put his hand on my shoulder and said, 'Thou knowst, Harold lad, thou must put up with it. Thy faither's got a difficult job on today and I can see his point. He conna ask others to volunteer to go down to a gob fire and let his own lad off wom to safety.' My ride up the shaft had not proved as exciting and full of relief as I had anticipated, for I knew now that I should soon be riding down again.

As usual the bright daylight made me screw up my eyes and, as I stepped from the cage, the first person I saw was my own brother, Harry. He joined me, took my lamp and handed it in at the lamphouse. Turning to him I said, 'If I'd been one rope sooner, I should not have seen my father.'

My brother gave me half a smile, touched my arm and replied, 'It wouldna have made a bit of difference, he towd me to wait and catch thee.

I dunna like the look of those ambulances; thy faither's been asking for volunteers to go and help seal off a gob fire or something. I've heard a chap talking of rescue apparatus, but I anna seen anything of the kind. When we get home, better not mention any of this to thy mother, it wunna do any good.'

'If my father had known the kind of morning we have had down Keele dip this morning he may not have ordered me back into the pit.'

'Oh, thou knowst better than that; he's never favoured any of the others in our family and he wunna favour thee. Thou must do what he says and make th' best of it.'

'Two lamps, two lamps, I dunna like th' sound of that. What the hell do I want two lamps for?'

Enquiring faces were at doors and windows as we walked down through the village. My brother stopped and touched my arm, 'Nagh, dunna forget, not a word to thy mother about the pit–bonk.'

'Well, thank God thou art back,' blurted out my mother with enquiring looks.

145

'I have got to go back to my father in South East.'

'What, got to go back, back down the pit where ther's a fire?'

'Yes, he grabbed me as I was getting on to the cage.'

'So thou hast seen thy father. What's he thinking about taking a lad into it? Oh, so he's already down th' is he. It anna taken them long to get *him* down into it. Now, who didst see on th' bank; were there groups of men standing?'

My brother's eyes flashed across to me, we had to be careful what we said for my father had sealed off gob fires before; my mother could filter evidence. 'Oh,' replied my brother, 'He did speak to some of them, I think he asked for volunteers, wanted help with bricking–off.'

'Bricking, bricking–off, doest say. So there *is* a fire then. Harold what didst *thou* hear, is it spreading doest know?'

'I know nothing except that I have heard the words gob fire. It is in South East.'

'Thy dinner is ready and in th' oven, get thy dirt off, thou't feel better.'

My brother produced a larger water bottle and soon we were back on the pit–bank. More men had gathered; we were relieved not to see evidence of rescue apparatus, but I pointed to one group of men and said, 'Look Harry, those men by the blacksmith's shop. Five of that lot are rescue men, I know that. Those five are all Silverdale men.'

There were two groups waiting to go down. It did my heart good to see them standing there for they were all volunteers. I felt proud that these men had all turned up to 'see if they could do anything to help'. Arthur Ricket, the man who had bellowed at me at the bottom of Keele dip, was one of the rescue men. Over in another corner talking to a pit–mechanic was Elias Green who had reported me and had me fined. They were all there, ready to 'go down' if they were required. All the resentment I had ever felt toward these men now dissolved in one moment of admiration for them. No matter what marks had been made on the veneer of my work–a–day life in this pit, those marks were now disappearing for I was now observing the *real* quality, the real strength of the timber itself. Tradition had taught me that when the test comes, miners are never found wanting in self–sacrifice. And here it was before my eyes.

As my brother carried my two lamps and we walked toward the cage, the word 'goodness' again came to my mind. I turned to my

146

brother and said, 'Harry, look at these men, every one willing to go down this pit to the trouble, not knowing what they will find, all volunteers. And I had the nerve to resent it when they pulled the rein a bit tight and put me in my place as a young pit–lad, learning the trade.' He nodded his approval but his mind was not on what I was saying; he moved very close to me as we stood there in a group waiting for the cage to appear.

Two young pit–men joked about the situation and their loud laughter seemed out of place, but I suspected that it was a cover to hide their inner apprehension. One older man leaned over, touched my shoulder and said quietly, 'Well Harold, thy father is a bugger for th' rules and regulations, but I'd rather be going down to this gob fire with *him* there than with anyone else. I used to hear men talking about him up at Number Six when he was working there with my own father and coal–mining was a different tale there than it is here.' Again my mental discomfort was eased by this man's remarks.

The cage suddenly lurched up before our eyes. My brother took my arm and with a worried look said softly, 'Careful kid, I hope it is owraight for thee.' This was the best he could do to express his concern, but his look was more eloquent than his words and I never forgot the expression in his eyes as the cage plunged downwards into the darkness.

There was complete silence on the cage. It was a very slow and quiet ride down. Somehow I felt that these volunteers were aware of the danger and silently hoped that those responsible for the decisions would issue the correct orders and instructions. In the pit–bottom stood waggons containing bricks, cement and other building materials; that was a good sign, for it meant that bricking—off had started. As we made our way to The Wall the two young men who had joked so freely on the surface now asked questions as to what a gob fire actually was. An older miner explained that coal and sometimes timber found its way into the space where coal had been extracted; that area was called the gob. When the roof came down and pressure was created, then internal spontaneous combustion sometimes occurred and the coal and timber began to smoulder and eventually break into open flame. This I had already learned from books and at night school.

At The Wall a man stopped us. 'I've been sent by George Brown to tell you to remain here.' Immediately I asked, 'Where is th' fire then, and is my father alright?'

'It is in th' owd gob near Hughes's dip. Thy father has not been sick, but he keeps holding his hands on his stomach. I dunna like th' look of those stretchers they have taken up there, I hope they dunna need them.' I knew that his holding his stomach was due to his everlasting digestion trouble; it caused him great pain. Within minutes two men were escorted outbye, both were vomiting and the sound of it sickened me. The man who had been sent to meet us, pointed out, 'You see what happens. There is gas up there and just one whiff of it and it goes down on to your guts. I had some experience with gob fire at Number Six and I dunna want owt to do with that muck again.'

Another light appeared; it was Mr Tom Burgess on his way outbye. He came over, dropped on to his knee and said, 'Now lads, there is some danger up there, but those men in charge have the answer if we keep our heads and do just as we are told; several of those men have had experience in this kind of trouble. If we can lay bricks fast enough we shall win. Just be patient, there is a good team up there. Just one more thing, the General Manager wants you to know that we are grateful for your voluntary help, I assure you that you will not be forgotten. Off you go up there and report to George Brown. Harold, you wait a minute, I want a word with thee. Thy father and the rest of that team up there are doing a good job; he knows how to keep his head when there's trouble. God alone knows what pits would do without men of experience. We are lucky to have men like that in this pit. Jos Jones is another and there are a dozen others like him, all rich in experience. I am grateful I can tell thee for having them around me today.'

Another man arrived from the gob, he drew Burgess aside out of my earshot; both men seemed to become somewhat agitated. Soon a man was carried out on a stretcher. Burgess turned to me and said softly, 'See what I mean Harold, a man's stomach cannot take pit gas.' Burgess left and accompanied the sick man on the stretcher.

Another light came from the gob and a familiar voice spoke, 'That you Harold?'

'Yes, Mr Deakin. How is my father getting on up there?'

'Oh, getting on with it quietly. Have you brought water for him? I think his stomach is causing him some discomfort.'

'Yes, there's a large bottle of water and his snapping in that manhole.'

'I've come out for a breather, Harold; it is my turn. I have not had it down inside me yet and I hope I can remain clear, but, oh dear, those who have had to go out were very sick.'

'Is ten minutes long enough for a man to recuperate?'

'It is according to how deeply you have taken the stuff down with your breathing. Your father wants you to go up there and give a hand and will you take the water? Walk slowly, Harold.'

'Righto, now I shall see myself what goes on.'

I walked slowly toward the place where the pit–lamps were glimmering and dark figures were moving about. There was no shouting, no light–hearted banter, no jocularity which one finds when miners work side by side; if they are not cursing the hard seam or struggling to set timber in an awkward place, they are telling colourful stories, or indulging in any kind of verbal exchange which serves to relieve the dull monotony and hardship.

My father arrived and lowered himself down; I noted his heavy breathing. He took a mouthful of water, rinsed it round and spat it out again. 'See lad, dunna take anything down into thy guts if thou canst help it.'

'How do you feel yourself father. Do you want any snapping?'

'Oh dear, nay Harold, where hast put it.' I told him where it was. 'Nagh, I must get out there for a breather myself. Here, you go up and relieve Bob Kimber on his waggoning; he will tell thee what to do. Now I must warn thee, when thou art right up on th' gob, take in short breaths. That kind of breathing will help thee to keep the muck off thy stomach. Keep thy voice down if thou hast anything to say, conserve thy energy as much as possible. The rubbish is still about, but if care is taken it can be kept away from thy guts.'

'What about the fire?' I asked in a whisper.

'Ah lad, it is still there, but we are smothering it a bit at a time. Another two hours like this and we should be able to stop anything coming towards us. Dunna forget now, ten minutes and then back thou comest into this main road.'

'Alright, I'll go and see what I can do, I hope that *you* will be alright.' He turned to go, but I glanced round at him and I saw that he was watching me; it was another of those looks which I was never able to translate. I fancied he was silently apologising for having to have me down there.

I took over from Bob Kimber who made a gesture that he was glad to be going out and quite characteristically, he joked, giving me a nudge, 'Harold, what about a good deep breath of that Spring Bank air thou't always boasting about?' What a tonic this was to me, a sense of humour at a time like this. We were in the main road where I had seen the two rows of ghosts on the night of the pumping; they were still there. Now I was witnessing a fitting atmosphere for my imaginary ghosts for again there was apprehension. No–one spoke in anything but a whisper; no shouting of orders, everyone exercising economy of effort, all taking in short breaths. Several officials were all busy in a concerted effort to shut out air from the smouldering fire, everyone seemed to know what to do. There was a kind of deep religious application—every man with his mind on one thing, the sealing off of this fire. Officials and workers all side by side, no–one assuming superior position or authority.

Soon I was on my way out for my breath of so–called air. The remote places of a pit often create a fantastic atmosphere, but now I was faced with something of another order. The muffled voices, muted further still by the curtains of brattice–cloth hung at intervals to direct air away from the gob, the quiet, slow movements of the men with closed mouth all the time, conversation in a whisper; all this created a sense of awe and mystery. There was an eerie silence, men seemed like creatures of another civilisation as they crawled and stooped about; my imagination gave me the impression that some great conspiracy was afoot. We were all striving to accomplish the task of defeating this fire and we were in a silent world of fear and darkness; and yet people up on the surface went about their business, enjoyed themselves, then slept in their beds quite ignorant of the drama being enacted deep below them in these dark caverns. All this in order that heat and energy could be brought to the domestic hearth and to the great power–plants of our land. How little people on the surface knew of life underground.

Our work went on with dull monotony, but with no complaint. The task was of the highest urgency. Men spat, coughed, vomited and some were taken out after reaching the end of their endurance. I saw one official after another take his lamp, turn down the flame and examine the conditions near the roof. I marvelled at the complete co–ordination of this gallant team of miners. No hurry, no panic; there was an indescribable dimension present. This created a deep impression on my mind and I am convinced that it is something which only miners understand. There was a quiet dignity everywhere with every man respecting his mate; self–sacrifice seemed to exist all around us; abstract qualities of life yet becoming physically articulate by the dedicated response of these brave men to that which threatened life and a means of livelihood. I had gone out to The Wall on orders from my father. Sitting alone in a man–hole I murmured almost audibly, 'No–one here has known which would win, gob fire or human effort, but there has been perfect co–operation, patience, harmony and even the occasional excursion into humour. No–one has witnessed it except the participants deep down in the earth; there will be no applause. These thoughts came to me as I took my ration of air, but with the thoughts were mingled the impressions I had received on the night of thirling with my father in Spencroft.

It was now evening and I observed an electric light coming from the direction of the pit. It was Mr Tom Burgess again returning to the scene.

'Harold, thou still down this pit? Has thy father been sick?'

'No, but he has had a lot of his usual stomach pains.'

'I'm just going to see him again, just in case I can be of use. How long since anyone had to give up?'

'Not for the last hour. That's a good sign I suppose?'

'Ah it is, thank God for it. Look Harold, thou hast been down this pit long enough. I'll ask thy father if he can spare thee.'

'Thank you Mr Burgess, I do not particularly want to stay down here, but if I am still of use I'm willing to stay. I shall be glad to get to my bed though.'

'I'll bet thou wut. Just stay here, thy father should be proud of thee.'

When Burgess did not return I went inbye and found Burgess and my father with lamp–flame turned down and watching carefully for evidence. Both men turned toward me and there was a whispered exchange of words. I heard Burgess say, 'But th' lad's been down th' pit long enough George, couldstna spare him and let him get to his bed?'

'If you order him out of the pit, then he must go, but I shanna send him home whilst these other men are still at work.'

Burgess turned to me and said, 'Off thou goest home lad, thou wutna have to be ashamed.'

151

I turned to my father, 'How long before the job is finished? Is everything under control?'

'That's something I can't say with certainty just yet. The next hour will tell, but we are encouraged by the situation. Tell thy mother that I conna say what time I shall be at home. Hast much water left?'

'Yes, about half the bottle is still unused. I'll take the snapping back with me then.' Hesitating, I gave him one last look, then turned to go but in my mind I asked the questions, 'Has he been successful?' 'What time shall I see him at home?' 'Shall I see him alive again?'

I returned to The Wall, made sure his water was in order, and looked back to the twinkle of the lights where men were resting. Suspicion crept into my mind that the bit of whispered conversation between Burgess and my father was a conspiracy to get me out of the pit for this final stage; were they getting me away from the real danger? On my way to the pit the words repeated themselves in my mind, 'The next hour will be the test.'

Another group of volunteers approached and as they came near one shouted out, 'Hello Harold, your father alright?' It was a member of our Sunday School. Another one touched my arm and asked how things were going; it was Harry Cleer who had taught me 'a bit about th' ripping side of pit–work.' The pleasure of this generous enquiry about my father was short–lived for as I reached the lights of pit–bottom I saw a group of men sitting in the entrance to Spencroft district. Seeing an electric light there I wondered at once, 'Is this a rescue brigade team?' I had recognised one man as a rescue member.

Charlie Lightfoot was the pumpman on duty; he shouted to me as I passed, 'Off home then Harold, hast had enough of it?'

'Yes Charlie, I've had enough alright. I've been sent home by the Under–manager.'

Three other men stood idly by the pumphouse. Only one of them was familiar to me; he was a man of the stature of Abe Deakin both in ability and character. He called out to me as I made for the cage, 'Hello there, young Brown. How is thy father getting on down there, is Abe still with him?' This further concern for my father pleased me. I replied, 'No, I think Mr Deakin has knocked off.'

I was alone in the cage going up and my mind went to the men I had recognised that day, all ready and willing to go down into the danger; Arthur Ricket, Elias Green, Harry Cleer and all the others. I felt more than a little ashamed of my early attitude and wondered if perhaps I had been a little too critical even of those who had humiliated me on the occasion when I received that disgusting initiation. I now realised that these men too were miners, that they swung every day at the end of a pit–rope in order to provide for their families. As I unlatched our entry–door I turned back and looked across at the pit. I tried to estimate the direction of South East district, then the approximate depth through the earth to the caverns where those men were still grappling with that vital problem. Before my mind I visualised them all hard at work and I gave them all a thought, 'Let us hope that they soon will have it under control.' It was a heartfelt plea and with it I turned to the entry–door again.

Movement at the other end of the entry meant that my mother was still up waiting. Her serious countenance asked the obvious question. Her only words were, 'Well lad?' I merely nodded as I moved to take off my clogs.

'Is he really alright then?'

'He seems alright, but a lot of his stomach pains.'

'Goodness gracious, his tablets, why didn't I think of them when you came back for your dinner?'

A cup of tea was produced in a trice and, not wanting food, I was soon in bed for I had tried to avoid answering too many questions in case I let out anything which may cause my mother to worry.

Next morning I was awakened by my brother who gave me a nudge, 'Had a good sleep kid?'

'Father, is he in bed then?'

'No, he's up. He didna get in till after one o'clock, so mother says.'

'How about th' pit, is th' gob sealed off?'

'Ah, I think so; I only know what thou hadst towd mother last night.'

Hurriedly I went downstair and saw my father sitting gazing into the fire. He seemed to be in a pensive mood. I nodded my head in the direction of the pit. 'Everything alright?' I asked, hardly daring to hear his reply.

'Ah lad, we managed it. We sealed it off, bricked it up completely. I'm sorry thou hadst to be there, but I couldna ask for volunteers and not take thee down.'

'It's alright,' I said, with some shyness. 'Now that it is over, I'm glad that I have had the experience. How many taken out?'

'Five I think, but we all got some of the rubbish down on our guts.

153

Didst thee get any inside thee?'

'Oh, a whiff or two, I felt a bit light–headed once or twice, that's all.'

'Hast slept well after such a long shift?'

'Like a log, thank you. Like you yesterday down there, I didn't want any snapping. It has been an experience I shall never forget and I *did* have the wind up I can tell you. If the danger is over, it has been worth it, but what an encouragement to see all those men willing to volunteer.'

'Ah lad, thou't learn as thou getst older that miners dunna shirk their duty when there is danger about, especially if other men are in danger. Thou't learn lad, thou't learn in time.' I fired a succession of questions at him on the subject of gob fire, but it was the gas which interested me most.

'I was not *sure* what the conditions of this fire were. It was Jos Jones who first reported the presence of gas. We both agreed to keep a close watch on it. The Gaffer told us that he would leave us to supervise and we made daily checks. When it did break out we knew that we had to do something quickly and it was a good thing that the officials on duty that morning knew how to set about things. There was no time to lose and we all took every precaution. Remember that, never leave anything to chance, there must be *no neglect in thy pit–work.*'

That afternoon I took a walk round Keele, and sat on the step of Keele Parish Church, but relaxation of mind did not come. My thoughts remained entrenched in the events of the previous day. Suddenly I became aware of Keele Village School and this revived my anticipation of Fred Askey and that engine. Fred had told me a lot about his village school and I now realised that I had to prepare myself for my tuition. I jumped to my feet and returned home only to find my father still sitting before the fire.

'What's the matter father, are you alright? Ought you not to be in bed getting ready for your night–shift?'

'I shanna go tonight,' my father replied dryly. 'That's all arranged. I did a very long shift yesterday and I'm feeling a bit off–side with all that gob fire rubbish still hanging around me. Hast had a good walk then?' With these remarks from my father I could not intrude further upon his patience to ask questions about Keele dip engine.

One night my father called me into the front room and gave me a serious lecture on my new responsibilities as engineman. 'Thy power extends from one end of thy rope to the other.' I seemed to recognise that he was now treating me as an equal, as though I had reached a stage in his estimation when I qualified as a pitman. This filled me with pride and I knew now that I could tell anyone that I had been appointed to perform this important job of work.

Les Bebbington had not crossed my path in the pit for several months, but one day we met in Back Lane. With a smile he asked, 'Ah–do Harold, how art getting on, still hooking bottom of Keele?'

'Yes, but I am to go on to Keele dip engine.'

'Well done, kid, thaiyt bey owraight Harold; that hooking–up has taught thee every pair of rails in Keele dip and everything else about th' dip too.'

'Thanks Les. I shall want a bit of luck for a week or two.'

'Good luck Harold, thaiyt do it I know, thou wast my first pupil in th' pit doest remember?'

10 Liberty

Walking up to the pit in my new boots I became conscious that they were a symbol of the fact that I was stepping up in the world again. My mind travelled back to the day when I was obliged to put aside my shop boots and put on clogs to go and work down the pit for the sake of the extra wages. I waited in the enginehouse for the arrival of Fred Askey. The engine was housed up in the roof. I watched the men walking underneath me and silently felt grateful that I, like them, was not having to walk through the long, low, narrow roads to get to my work. My thoughts and feelings were very mixed indeed.

At length Askey's head appeared at floor level as he climbed the ladder. There it was, his usual soft smile and I would have given anything to have known the quality of his thoughts at that moment. He tested the controls right away and stood on the brake to feel the resistance; he turned to me and said, 'Harold, for the time being, just watch.'

After three journeys had been landed I was allowed to hold my hand on the brake lever to 'get the feel of the weight of the journey.' Although it was a foot brake the empties had to be braked by hand in order to exercise sensitivity of control. Already I knew that I was going to enjoy that part of my work. I never took my eyes off Askey; his concentration and reaction to the bell–signals were an inspiration to me. He said little for the engine hardly stopped. After over an hour he leaned over the safety guard–rail and said smilingly, 'Think we shall make an engineman of you Harold?'

'I hope so, Fred, I'll do my best. The rest is up to you.'

I had already observed that even this experienced man became tense as the journey entered that part of the dip with an extremely steep gradient, that part known as The Steer. Now came the teaching proper. I was now to be taught how to pick up the journey on this steep section, the journey of heavy loads, coming up.

'Sometimes, Harold, your journey may have to come to a standstill on the steer. If someone in the dip knocks you "hold" on the bell, you *must* stop. Indeed, sometimes, your main switch will drop out on account of the journey having more than the prescribed number of loads of dirt on it. Dirt is heavier than coal as you no doubt have learned and there must not be too much of that on a journey. So, I am going to stop in the steer and let you feel the weight of the journey under your foot; you *must* stand on your brake under these circumstances.' So the instruction went on all day, one element of technique after another; one different aspect of the work after another. He never became ruffled and it was a pleasure to me to hear the softness in his voice and gentle manner as he said, 'Notice, Harold, the very gentle way I pick up the slack rope at the bottom of Keele dip when starting up. You've seen it all happen down there when you have been hooking.' More and more experience in stopping and starting on the steer was entrusted to me and by the end of the shift I had brought up six journeys myself. By snapping–time I had come to the conclusion that one thing alone would establish my efficiency in this work, I must become expert at picking up the journey on the steer from dead–stop. I had already noted how Askey's engine–driving was just like his character. There was no banging and clanging of the machinery as he engaged gear; everything was done with such ease and without fuss.

Fred Askey had persuaded me to eat my snapping as the engine was running and he too munched his sandwiches as the journey came up, but always with his hand on the controller. Snapping–time came and Askey extended me the courtesy of pointing to his box, indicating that he was offering me a seat. 'Harold I asked you to eat your snapping, for it is necessary for you to concentrate on what I am going to say to you. You know this dip like the back of your hand. A long period of hooking–up has made you familiar with every aspect of the dip; low sections of roof, flat sections in gradient; you *do* literally know every pair of rails. Also you have worked on this landing here in pit–bottom and you are familiar with that part of the work. Bell–signals are as familiar to you as your own name, so there is just the engine now for you to learn.

'The most important thing about this machine is its noises. You will learn to recognise its various tones. The noise will tell you if the journey is overloaded; it will tell you where the journey is in the dip although the laps of rope on your drum will also tell you that. Remember now, always, first journey down or up, always gently for a hundred yards or so. Do not assume that the Road–doggy has cleared

157

the track for certain. Assume yourself some minor obstruction on the rails somewhere for your first journey. If you are in any doubt whether or not you have received the "dip clear of men" signal from the bottom, just ring down again two–two's and ask for the signal again. Now it is signals that are also very important and I want you to pay careful attention now.' Here he took a deep drink from his bottle and nodded for me to do the same. Again he went on, 'You know the signals, Harold, but the manner in which they are given is as important as the manner in which you take up your slack rope at the bottom. You cannot afford to have a bad temper on this job, you must be calm and level–headed *all the time*. Lives are in your hands, so leave your temper at home. You will be inclined to give your bell–signals according to the mood you are in. Leave your moods at home also and come to this engine composed and with a quiet mind. Never allow the speed or rhythm of your signals convey the idea of hostility or aggression, for this will snowball and return on to you own head like a boomerang. You know this from your own experience. You can't change the tone of the bell for that is static, but you can control the speed and the rhythm of each signal. When you ring down to the Hooker–up never let it suggest that you are saying, "I'm ready to run in, come on now, let's hear from you, don't keep me waiting here." That kind of signalling does not help anyone; never give in to that kind of temptation. Your signals should "ask permission" rather than "demand the right". Give your signals with politeness in your mind. Even in pit–work Harold, you will find that it will repay you a thousand times if you are polite on your giving of signals. After all, they are your only means of communication with your mates at dip–bottom and those working in the dip who are authorised to stop your journey. One other point, if you stall your journey on the flats by not allowing enough speed to get it over those stretches of flat road, do not worry. Engage gear, draw her back to a gradient, then let her go down again. The journey is *yours* once it's in the dip, yours till you get an authorised signal.' I took in a deep breath for it had been quite a speech. Also I noticed the way Askey stressed the word 'polite' and even as he was talking to me, I recalled in my mind the number of men who had been polite to me since I started working down this very pit.

At the end of this first shift Askey relaxed on his box and asked, 'Well Harold, how do you feel after your first day with this engine?'

'Fred, I feel grateful to you for the trouble you have taken. I had no idea that anyone would take such pains to guide me over these first hours. Thank you very much indeed and for your sound advice. Do you think I shall make it?'

'But you have made it, haven't you. You have brought journeys up and landed them. You are a bit young, see, only just turned seventeen. But you'll make it, you will make an engineman alright for you have control. Now here is a sheet of paper, study it and memorise the points I have already touched on at snapping–time.'

My body had been rested that first day at the engine and I was not physically tired, but my mind was pre–occupied. After my meal I settled into the rocking–chair to study Askey's notes. He had described it as a creed for enginemen and that is exactly what it was. When working nights, make sure that you get your sleep in *first,* immediately your breakfast is over; get your sleep in before the street hawkers come round yelling their heads off. If you neglect your sleep, you may doze off at work when your engine is in motion. You can't afford to do that. It is a serious offence to be caught sleeping anywhere in the pit, but asleep at an engine, why, they'd hang you. His notes went on and on covering all the points he had made with me during snapping–time and my mind remained on the word 'polite'. I wondered how many men would have taken the trouble to write out these notes just to make sure that a young pupil would have the facility to turn his eye and mind to them and memorise them. I began to feel that some form of compensation was appearing in my favour. Grim as the gob fire experience had been, the final result, plus this very gratifying day with Fred Askey, seemed like a fresh lustre creeping into my existence among these mining men.

I decided that evening to turn my back on my normal study routine and consider Askey's advice, for my mind would not let go of it. As usual I turned to Back Lane and the fragrance of Davenport's fields and woods. Over and over it went in my mind, and, just as I had repeated the various layers in the earth's strata, during my first lessons in geology, so my mind now had a task of another order and I knew I just *had* to memorise Askey's notes—'Get your sleep in first; don't lose your temper; my job must come first; I must maintain control at all times; men's lives are in my hands; don't abuse the power under your hand; always politeness with signals; etc.; etc.'

My father asked for a recital of the day's events at the engine, but I

159

carefully avoided Askey's reference to 'politeness'. I told him everything else, but thought it best to keep that part of the day's work to myself and give it some thought. 'Was this too soft an expression to be associated with miners and coal–getting?' I was willing to agree fully with Fred Askey, accept his advice and rely upon his directions. I read and read into his carefully worded notes which suggested that he could teach me the manual skills required in engine–driving, but to be a good engineman one had also to understand men and to be aware of responsibility towards them. Letting down the journey of empties at great speed, hauling up the heavy journey of loads with the attendant stress and strain on the rope required a very steady hand and a cool, undisturbed mind. Quite often an engineman graduated to become winding–engineman, in charge of taking men up and down the shaft. This was *the* most responsible job in that department of a pit. I said to myself, 'No, I'll stick firmly to Askey's advice. Politeness on the bells is quite in order. No–one can interpret this advice as soft, for Fred Askey is as manly as any hardened collier at the face. There was nothing incongruous about advice to a young man handling for the first time a powerful machine, that he should maintain a diligent, serious and responsible attitude of mind in every aspect of his work.

When Askey considered that I was sufficiently in control, he spent some part of each day in various parts of the dip. He would stop my journey at any point he pleased including The Steer and then he would return to me at snapping–time and report to me his own opinions on my ability. I felt very flattered.

'They are pleased with your driving at dip–bottom Harold.'

'Thank you Fred. By the way, who *is* the new Hooker–up?'

'Believe it or not, it is your old school–pal, Les Bebbington.'

'Well–done Les. No news could have given me more pleasure than that. Les and I working together again. How well things have turned out.'

Difference in pitch had been part of my studies in music, but now I was listening to other sounds as my engine groaned out its mechanical theme shift by shift and let me know over which gradient the journey was passing. Sound was now becoming part of my very existence but in a more rugged context than in the refined nuances of Bach, Beethoven, Mozart and Chopin. Dirt now to me was as important as the coal itself; extra dirt on the journey brought complaint from the engine and a warning to me to be ready for the main switch to drop out at the steer.

The months went by without serious mishap in my work and I was given to understand by Mr Harvey Rogerson and others that I was giving satisfaction. The sudden change from the type of work which demanded physical strength and endurance to an occupation which demanded less muscular power, but more mental alertness and concentration seemed to create the right conditions for something very inherent in my nature. It had always been there, but physical demands had crushed it. Now, more than ever, I became conscious of the urge to sit quietly and think. Along with many other serious–minded people I needed the facilities which would enable me to withdraw from the veneer which we called 'everyday life' in order to enter into an atmosphere of quietude. My job at the engine provided these conditions, but I had been warned by Askey that I should have to guard against over–indulgence in this mental preoccupation; he had made a very strong point, *'Your mind on your job* every moment your engine is in motion.' Even though I was now fully established as an efficient engineman, Fred Askey remained constantly in touch with me. I was still virtually his pupil and he paid me the compliment of keeping me within his orbit of duty and I accepted his diligence with grateful mind.

After about a year as engineman Fred Askey climbed up into the enginehouse one day to eat his snapping. Having finished his food, he turned and said, 'Harold, have you ever entertained the thought of becoming a winding–engineman? You are teetotal, you do not smoke and you live a clean, steady life. You are good at this job with perfect control, you have all the attributes. Give it a thought Harold for it has been mentioned in other quarters.'

'Thanks Fred, but have you not heard the rumours of yet another strike, the strike to end all strikes, so they say? I shudder at the thought. Harvey Rogerson says that the coal industry is dying. So, I think it is time to look at the future and try and estimate what the chances are of our being able to get a living in the pit at all. Also, I am supposed to be studying with the object of possibly getting my higher papers with the aim of Pit–manager in the future. Goodness, I've done plenty of study already in that direction.'

Askey became excited and stood up, 'Harold, don't you see, just as a bandmaster or an orchestral conductor has to become familiar with the technique of all the instruments under his baton in order to command respect and help a player with his performance, so a Pit–manager has to equip himself with every branch of pit–work. A man with First Class papers offering himself as General Manager would stand more chance if he could show that he was also an experienced winding–engineman. Do you get the point Harold, lad? It would add to your final qualification as well as provide you with the time at weekends for your study while you were on winding. There's not much winding at weekend and on Bank Holidays.'

I smiled at Fred Askey's remarks and appreciated his compliment, but after a few moments of thought I told him my inner feelings lay in another direction. 'Fred, I've thought a lot about it lately and I really am beginning to think of the possibility of getting out of the pit. I know Fred that there is not a lot of work to do when I'm at work at weekends, but every other weekend I'm expected to be here. Repair work in the dip requires an engineman; I can't refuse for my father has difficulty getting men to work weekends because they lose this subsistence allowance. You can't blame the men, but, you see, that is what happens to me each time I work a Saturday night or a Sunday. The subsistence is valueless to me. I started work nearly four years ago in this pit at thirty five shillings a week; I am not earning two pounds now on a full week. That's how wages have decreased. With no prospects in the future, why stay in such a job?'

Six months later Harvey Rogerson came to me, 'Harold, short–time does not affect you on your job but have you not observed how the periods of short–time are becoming longer and more frequent? Things are looking bleak in our trade.'

'Yes Mr Rogerson I have seen the long lines of trucks filled with coal which no–one wants and I've heard the shouting again about a strike. I've been to a Union meeting and heard the same arguments and threats from the speakers. I can see the same worried looks on the faces of the women and the small tradespeople. Desperation is creeping in again and I begin to feel afraid. I could not put up with that humiliation and suffering again.' Rogerson smiled and said in a fatherly sort of way, 'Harold, lad, you are worth better than the pit is offering; think seriously about it.'

That night I talked with my father about the possibility of my trying to get out of the pit. He said little till he had finished his meal, then, pushing away his plate he said, 'Hast thought about it, Siree?

Dost *really* know what thou't talking about? Get out of th' pit, get out of th' pit? Just mention that to any of these men who have worked underground all their lives. See what they will say to thee about it. What about thy studies at night school, what about all thy effort? What hast been studying for if it inna to get thy papers, go on to th' face and qualify for shot–lighting?'

I looked straight into his eyes; he was worthy of all my confidence and esteem. I told him of Fred Askey's suggestion that I should consider the possibility of my one day applying to be trained as winding–engineman. 'I could then study and possibly take some course in scholarship; this would be like a second string to my bow, as it were.'

'Scholarship dost say? So thou't still hankering after learning. Thou hast never been any different, always a book in thy fist. Thou art a pit–mon now and th' only books thou wantst are those books of mine on practical mining.'

My father was right, for the next day the subject of 'getting out of the pit' was discussed at snapping–time. An older miner laughed and said, 'Out of the pit, dunna talk rubbish Harold, once a miner, always a miner. Another thing Harold, thou'st forgotten, there are now two men waiting for every job which becomes vacant. What would thy faither say, a good enginemon with a nice safe job, getting out of th' pit? He'd kick thy arse, old as thou art.'

These remarks did not add to my zest to make an attempt to cut clear of the pit before the storm burst. I indulged from time to time in some retrospection and tried to assess the sum total of the knowledge I had gained by means of study and in what field I could offer it as a qualification in my application for other work. I was certain that each time my type of work had changed in the pit, each fresh experience I had encountered, had sharpened my wits a little.

I attended another Union meeting and there were speeches to the effect that the Railwaymen, the Dockers, Transportmen and other major unions had pledged their support for the miners. The whole working population had now recognised that the miners were grossly underpaid and a general strike seemed to be inevitable in the not too distant future. Unemployment was rife and everywhere I saw hardship again, just as I had seen it in 1921. So I began my trek around the Potteries towns to factories, engine sheds, warehouses, and even grocery stores where I offered my experience as a boy of eleven to fourteen as a recommendation, an argument that I could do the work if given the opportunity. My mother noted my change of habit; instead of my going straight to bed after my night–shift, I changed into better clothes and left the house, thus ignoring Fred Askey's plea that I should get my sleep in first when on nights. My mother's looks betrayed her concern. 'Where hast been all the morning? Why art going out instead of going off to thy bed?'

I tried to evade her searching questions and looks, 'Oh, I thought I'd get some fresh–air. Driving that engine affords me no opportunity for exercise.'

'Ah, thou canst tell that to someone else. You dunna turn toward Newcastle and th' Potteries to get fresh air. It is to Keele and Madeley with they face towards Wales for fresh air. Is it a wench thou art meeting in the mornings?'

This question amused me for already I was paying my attention to a young lady who lived only a few yards from our house.

'No! It isn't a wench. I am doing my best to find a job out of the pit.'

My mother gasped, 'But you have such a good job up at Kents Lane pit.'

'What's the good of that if the pit is going to stop working?'

'Well! I don't know what your father will say, *you,* his son, leaving a job like engine–driving.'

'What can he say, what can he think? I'm only trying to make sure that there will be a bit of money coming into this house when the strike does come. Do you think that I *want* to leave my job just now? I enjoy my work; it gives me a chance to study and one day I may qualify to become a winding–engineman, even if I do not take my papers and aim higher. I know that my wages are very little higher than they were five years ago, and I have a clean job with advantages and prospects if things stay as they are, but things are not going to stay as they are, everyone knows that.'

I could see that it was a waste of time trying to make anyone see my point of view and the very day of this conversation with my mother I had observed the procession to the pawn shop, a sad procession which seemed to me to be getting longer and more pronounced. It was obvious to me that no–one knew just how deep the mark was which the last strike had made upon my mind, neither did they know of the echoes which kept recurring, echoes of those strikes during my early childhood. The marks were all still there. My search for other work went on for two months. Then, at last, news came of my application to our local branch of Swettenhams Limited, family grocers. The branch–manager informed me that he had persuaded his governor to give me a chance at a pound a week wages; he called at my home to see my mother. 'But Mr Higgins, I could not keep Harold on a pound a week.' 'Mrs Brown, we do not give our fully–trained young men much more than that. Will you give your lad the opportunity and I will finds means of getting more money for him when he has proved his worth?' Swettenhams had branches in the surrounding villages and were opening up new ground in the Potteries. It was a go–ahead firm and in keen competition with the Co–operative Societies. We were obliged to wait and see what my father's attitude would be to this.

Now occurred an episode in the pit which ranked equally in importance with the night of thirling, the gob fire and my promotion to engine–driver. One of the pits which had closed was situated only about six hundred yards from my own pit. The old shaft had been widened and sunk to the level of our own seams and now we had the benefit of a wide new shaft and headgear. A large new pit–bottom had been built with walls some twenty feet high, built of brick, and enough width to take five tracks of rails. It made me imagine sometimes that it was some great underground theatre or cathedral. Keele dip engine was brought down from the old pit–bottom and re–housed up in the roof of this new complex. A new link road connected the new pit–bottom to Keele dip, joining it near Plate Landing.

The Road–doggy in this new pit–bottom was Charlie Wade, a short, strongly–built man and a willing worker. He had recently married and when time allowed in coal–drawing he was subject to a certain amount of good–natured banter, leg–pulling and ribaldry. Charlie took all this in good part; he was too happy in his early married life to mind. One night as we all sat together eating our snapping Charlie volunteered some of the delights of married life. There seemed to be no end to the advantages as compared with the loneliness of those of us who were yet unmarried. Charlie's talk betrayed his happiness, but on this night our laughter suddenly ceased as Charlie adopted a more serious tone of voice. We all waited silently for Charlie to continue; there were five of us present. A soft smile spread over Charlie's countenance, it was no grin of sensuous satisfaction, it was more of an expression of deep inner happiness. He began to speak softly and clearly, such a contrast to his usual rugged manner and rough pit–language. Everyone present changed his attitude as if by instinct; that moment became imprinted on my mind for it did not belong to the realm of ribaldry. He looked round at us all, 'What have all you lot got when you arrive home from night–shift? You are all single. You have your bit of breakfast and off you go to bed. When I arrive home I have a loving wife and a warm embrace waiting at the door for me, hot water ready for me to get my dirt off, then a good hot breakfast. Look at you all, you know nothing of going home to a woman you love. Do you know, if I had to work a shift and a half every day of my life, if it took every penny I earned to keep my life as it is now, I would be happy to do it.'

One pit–bottom worker, not to be outdone stood up and announced, 'I'm getting married soon; we have got nine pounds saved up between us.' Even this naïve remark produced no laughter, we were all still under the influence of Charlie's serious attitude and his amazing speech. He had proudly declared that no matter what the cost, no matter how much effort may be necessary to keep the woman he married, he would be happy to do it. There was no laughter as we all rose to resume our work.

As I climbed the ladder to my engine I turned and said to one of the men who marshalled my journeys, 'What a tribute to a woman. My goodness, how proud he is of her.' A few minutes remained before I received a signal to start the engine, I looked down at the huge underground vault, this new pit–bottom and my mind relived the few moments during which Charlie Wade had made his serious declaration. In my imagination I said to myself, 'Charlie Wade has just endorsed in his simple words the vows of his marriage. Here down the pit, he has paid his wife the highest compliment.' Coal–drawing re–started and for the rest of that shift it was one of those periods when all went well for Keele dip, for my rope never stopped; we were very busy indeed.

As I let my journey of empties down silently I did hear a bell–signal on the South East bell that help was wanted and I turned to see Charlie Wade grab his lamp and make off to investigate the stoppage. At the end of the shift I made my way slowly to the cage in my usual manner. Soon I learned that Charlie Wade had been killed. I then recalled in my mind that I had seen a stretcher being carried into the South East district, but that was such a common occurrence that I did not give it a second thought. But, 'Charlie Wade killed, Charlie Wade dead.' Only a few hours earlier he had been so full of life and happiness. 'My God, my God,' I said under my breath, 'Charlie Wade has certainly paid the very highest price for coal.'

Mr Higgins, the branch—manager of Swettenhams Limited called again and emphasised that he would like me to start work as soon as possible in order that I may become accustomed to their method of booking and their credit system. There were now only two weeks before the busy Christmas trade.

'Have I got the job for sure, Mr Higgins; is there any question about it?'

'No, Harold, if you can start on Monday next.'

I pleaded with my mother and said that if she would make the sacrifice I would forfeit my six shillings a week pocket—money for six months. 'Well, I dunna know what thy father will say, I dunna know.'

I went straight to the pit—bank and asked Mr Burgess if I could have my liberty, leave the pit at the weekend.

'Liberty! Liberty! Did I hear thee say liberty?' He stepped forward toward me and I, fearing the worst, retreated. 'Damn me! Damn me!' shouted Burgess, 'What the hell next? I'll give thee liberty lad. Why, most men would give their right arm to have thy job. Thou hadst better get thyself away from this pit—bank before thou feelst my foot in thy backside. I'll have a word with thy father when he comes up the pit.'

I hurried away uttering my own complaint. 'Good God, now for it at home. And I'm doing all this only to make sure that there will be a bit of money coming into the house when the strike comes. How easily I could sit at my engine, read my books, earn my present wage and then do nothing when the trouble starts. But I *know* I'm doing the right thing.'

My father had recently changed from all nights to all days; he was now Day Overman. It was now Wednesday and I was on nights. I related to my mother what the attitude of Mr Burgess had been.

'Thou hadst better get off back to thy bed; thy father will not disturb thy rest. I'll have another talk to him and I'll do what I can to make him see th' sense of it.'

167

I did not sleep. I heard my father arrive, then a muttered conversation which developed by crescendo into louder voices. I decided to go down and face it. My father's features revealed that my fears were not groundless.

'Well, Siree,' he shouted at last. 'What did Mr Burgess say to thee? I wish he belted some sense into thee. Look! Hast thou never heard me in this house complaining about men asking for their liberty? Anna it got into thy thick skull yet how difficult it is for officials to replace skilled miners asking for their liberty? You conna replace a man at a day's notice no matter what job he is doing. Every miner has a contract—, two weeks notice on both sides. What dost think we have that rule for? If thou hadst th' sack, *thou* wouldst expect to be given two weeks notice. Thou wouldst expect the Management to keep their contract and give thee two weeks to find a job somewhere else. And here is my own lad, going to th' under–manager and asking for liberty. Thou artna too owd to be put in thy place and that's what thou wut get if I hear any more in this house about liberty. I conna stop thee leaving the pit if that's what thou wantst, but thou wut work thy notice like any other mon, two weeks to the day. Nagh, dost understand that?'

I did understand and I retreated for it was not an idle threat and I was not so long out of my schoolboy days to have forgotten the pain of my father's foot in my hind quarters with his full weight behind it. I left the house, knowing full well that my father was quite justified in his attitude; I had heard the same argument many a time from miners. But I mused as I walked aimlessly, 'Surely, surely, they can stretch a point when a young man has a chance to better himself?'

I had a problem on my hands and I turned my mind to whatever power there was in the Universe which might inspire me. It seemed to work for suddenly I thought of the man who drove my engine on the opposite shift. 'He loathes nights, I'll ask him to work my week of days.' I had no trouble with this and the following Sunday I arranged to begin a fortnight of nights. I had handed in my notice to the office on the day before without a word to my father, hardly daring even to mention the subject. My father was in a sulky mood and both he and my mother knew that hard work and effort would not deter me from my plan.

As I prepared for the pit on Sunday evening my father looked up from his paper and asked. 'What art doing, I thought thou wast days this next week?' I had my reply ready.

'I am, but I've changed with the other engineman.' No more was said and I worked that night and arrived home at 6.20 am on Monday morning. I washed, had breakfast and lay down for an hour before going to the shop at 8.30 am.

'Shall I be alright, shall I be able to reckon up quickly enough? What about the bacon—slicing machine?' Questions, questions, mixed with apprehension. I tried to pretend that I had not worked all night. As I put on my clean, new, white grocer's coat and apron, I felt the cleanliness, and already compared it with my pit—clothes, the wet, dirty shaft and the stink of the sump.

Within an hour I realised that things *had* changed somewhat since I had worked in a shop five years ago. Higgins was too busy to attend to me, so I was left to 'use my common sense and get on with it'. My mother's training, my own nature, or, whatever power it is which grants us our gifts at birth, or perhaps a combination of all these elements, I seemed to have been blessed with the ability to see a job which needs to be done and then get on with it. But I was also blessed with an elderly assistant at that shop. He was John Spode who kept me in his eye and gave me words of advice and encouragement. Everything was so strange after pit—work, but this saintly man helped me to overcome the more difficult tasks. Lifting the heavy sides of bacon up on to a high rail using a long pole which was greasy; this required a kind of skill which is acquired only with long practice. Taking the bone from the middle of a whole ham without bringing away even one scrap of meat, all this I had to learn to do. John Spode has sons of his own in pit—work and he was displaying the conduct I had experienced in the pit; John Spode was confirming my own observations that the high standard of comradeship underground does indeed seep through to the surface.

I arrived home that first night feeling very, very tired and Fred Askey's words buzzed in my ears, 'Never neglect your sleep when on nights Harold.' Working at the engine had rendered my muscles somewhat soft and now in my new work physical demands were being made upon me again. The third day brought strain and I was showing it. John Spode, bless him, knew just how I felt; he also knew that I was making this great effort to get out of the pit. Tiredness was my enemy and, with it, the inclination to doze off to sleep at my engine. By the end of the week, even my father relented as he observed my fatigue. As I left the house for the pit on the Friday night, my mother took my arm and pleaded, 'Now, remain on thy feet lad if you can. DO stay awake.' Mr Higgins also was not blind to the effect the long hours of work were having upon me and he had allowed me to go to the shop one hour late on two mornings. After shop–closing on Saturday I went straight to bed and remained there till it was time for me to go to the pit on Sunday night, getting up from bed only for my meals.

Now I was faced with my second week of doing two jobs. At the end of my Sunday night shift and a long day at the shop I felt very tired and my weekend in bed became neutralised. The next three days took full toll of my strength; my fatigue was now increasing by geometrical progression; it was showing not only in my countenence, but in my every step. My father came completely out of his sulky mood for he could see what was happening. As I left for the pit on the Thursday evening he followed me down the entry, held my coat–sleeve and said seriously, 'Nagh, dost hear Siree, thou wutna drop off to sleep wut? Remember the lives in thy hands when thy engine is running. Dunna go and spoil everything by dropping off to sleep. It's a grave offence thou knowst in th' pit. Stay on thy feet, try not to lean on anything. Why art taking such a lot of water?'

'I use it to give my face a cold rinse every now and then to try and remain fresh and alert.'

'Well, stay on thy feet now, dunna sit down at all.'

The next day, Friday, with only one more night to do at the pit, Mr Higgins asked, 'How are you getting on, Harold. Are you still driving that engine at the pit?'

'Yes, Mr Higgins, I have not missed a shift, but I find it hard keeping my eyes open. Am I proving of use to you here?'

'Yes you are, Mr Spode says that you will make it alright. You are a good worker; I'd heard about your family tradition for being good workers. No–one will be able to say that you have an idle bone in your body. Try to keep going and you can come in an hour late tomorrow again. That will help just a little.'

I suspected that he was beginning to doubt my ability to see it through. So I struggled through Friday at the shop and faced my final night at the pit. I was so tired that I could hardly place one foot in front of the other as I struggled up Back Lane to Park Road. The meal was set before me, but my tired body refused; I slumped across the settee and remembered nothing till I was aroused by angry voices.

'That lad inna going to that damned pit again. This is killing him. If he goes to that engine tonight he wunna remain awake and thou George Brown wut be to blame.' I opened my eyes and there stood my mother addressing my father; she appeared like some general doing battle. My father said nothing, his cheeks were white and he had a look of bitterness on him, he merely gazed down at me.

My mother then addressed my younger brother Harry. 'Here, Harry, get thysel' up to that pit and tell th' Night Overman that Harold wunna be at his engine tonight. Just say that thy brother has worked himself to a standstill and that it will be dangerous if he drives the engine tonight.'

I was too tired and worried to take in the significance of this situation, but throughout life the scene has reappeared before my mind's eye. There was my mother, mother of a large family, giving orders as to what should be said to an official at the pit. My father was, without question, master of our house, but my mother who had seen nothing but hard work and rough conditions, sometimes bordering on abject poverty, was now taking command, opposing her husband and defying the authority of a coalmine. She was doing all this merely to protect one of her children who was being threatened. She stood over me like a tigress and I fancied that she was saying within herself, 'Enough, enough.' My mother was known as a hard woman and she lavished little sympathy on any of us; life was too much of a battle for that. But she was quite typical of the women of North Staffordshire in those days; tough as nails to deal with the hard conditions of life and to protect her young ones if danger threatened, but she was as gentle as a lamb when compassion and help were needed at the home of some unfortunate neighbour. I had just witnessed a display of her true character and metal as she refused to allow me to be tried beyond my physical strength. My father left the room, but my mother turned to me and gave me a long look, a look which said without a word being spoken, 'I'm sorry for thee, lad, but I've put an end to thy trial.' That look lasted several seconds only, but to me it was an age, for, contained in that look there was a long, long period of history of evolution; the history of the development of a parent's inborn concern for offspring. I was not at that moment aware of the deeper sacrifice my mother was yet to make; my wages were going to be halved and this was going to make things difficult for my mother. There would be strain in the realm of house–keeping money, and my mother was

171

going to make this sacrifice 'for me'.

It took weeks for me to realise that I should not have to swing at the end of a pit–rope again. My financial loss brought a great strain but in my mind I felt assured that I had done the right thing and cut myself clear of the coal industry. My long shop hours gave me little time or energy for reading, study and piano practice, but I had the precious commodity of life around me, light and fresh air. Occasionally I made my way to Spring Bank and relived the happy hours I had spent there in earlier life. The walk in that direction took me up the steep hill leading to Silverdale Vicarage and to Keele. Each time I climbed this well–known hill I heard the clangour of the pit–bank activities, the cage banging down on the supporting legs, trucks of coal being shunted in the sidings. Once arrested in this way I always stopped, turned and looked back at the pit complex, the chimney, the headgear and the dirt–tips. A tense excitement seemed to run through my whole being as I murmured to myself with religious gratitude, 'I've escaped . . . I've escaped.'

One Sunday afternoon, soon after my change of occupation, I was putting on my coat before leaving the house for a short recreational walk. My father had not yet returned fully to 'normal exchanges with me since my dramatic departure from the pit. Nevertheless his attitude was becoming slightly warmer and I was surprised when he suggested that he accompany me.

Almost immediately after leaving the house, we encountered a small group of young children, some of them poorly clad and neglected in appearance. Two of them wore no shoes nor stockings and this arrested my father's attention. He stopped and looked at the children and I observed his concern. Smiling at us one child looked up and said, 'Hello, mister'. This pleased my father who returned the smile and warmly replied, 'Hello there, have you all been for a nice walk?'

We continued our walk slowly on a circuitous route. 'Let's walk round the Park, Harold, I have not done that for years,' said my father. I silently agreed but I was aware that something in that group of children had impressed him. Soon we found a place to sit on the green grass and gaze across in the direction of the top end of Silverdale village. After a few minutes silence my father remarked, 'Harold, didst notice those two urchins without footwear, nothing at all on their feet? I felt very sorry for the poor things'. I paused before replying for within me I was happy to see my father's sympathetic reaction to the unfortunate children.

'Yes father, I did notice them and I also observed your own interest in them.'

'Harold lad, I am sure that thou must remember that when you were young there were times when there was not enough to eat. There were a lot of mouths to feed and wages were very low and there was a lot of short-time. It was not easy for thy mother and me to see you little ones go hungry. But didst notice that thy mother always made sure that the pit-men had a good solid meal on th' table at the end of every shift, that was necessary to maintain a man's strength. We were ver poor with only just enough to pay for bare necessities but none of you were ever turned out of the house without footwear. You all wore clogs all the time except for Sunday School and Chapel but so did most other children and there was no shame in that; after all, your clogs sparkled when you had all cleaned them every morning with blacking, remember!? You had empty bellies sometimes but we tried to turn you out clean and tidy; it was hard scraping the money together to buy new clothes for you all when Sunday School Charity came round.' Silence followed and I suspected that my father was trying to apologise for the hungry days of our childhood. I remarked quietly, 'We are all aware of the sacrifice you both made for us.'

My father seemed anxious to go on talking, 'Those poor children in bare feet brought those hard times back to me. I used to afford myself only small pleasures, a game of bowls and a drink after the game. One cigarette a week and one cigar at Christmas which thy mother always gives to me as a Santa Claus present. There are not many children now without footwear. Thank God for those volunteers who have worked and fought to put an end to our seeing children in the streets in bare feet.'

Returning home I retired to our front-room and the rocking-chair where I put my head back, closed my eyes and I too thanked God for a mother who was an efficient house-keeper and a father who made sure of provision for our bodies before taking anything out of life for himself. Both were typical of the men and women who constituted our mining community.

Epilogue

Grim tenseness prevailed everywhere; the fear and the apprehensive mental climate could be felt in all corners as the whole nation waited for the strike to start. It came at last, but the nationwide support did not last long; the General Strike soon collapsed and the miners were left to fight on alone. The long weary battle ended again with utter defeat for the men. The coal industry suffered most as the nation as a whole plunged into industrial devastation and soon the country had declined so rapidly that it entered into a period which became known as 'The Industrial Depression'.

I had changed my occupation just in time and I put all my effort into my new work. The local Branch Manager kept his word and taught me how to obtain new business, my first area being the village of Keele itself. My wages increased, I was promoted, I became self–assured and proud to be making a major contribution to the family purse. But I could not escape the utter poverty and abject misery which was manifest on the faces of so many suffering people. Also, I knew that I was missing that unique relationship which had grown upon me among miners underground. People would come to me as we closed the shop on Saturday nights, 'Have you a few bacon bits over, or a few bones to go into the pot to make a meal tomorrow? It is Sunday, my husband is on the dole and there's nothing at all in the house.' My sensitive nature was suffering and I found myself putting an estimate on future prospects amid this dire poverty.

One wet day, as I trudged on my round through the dark streets of Stoke–on–Trent I was arrested by an advertisement which promised, among other things the prospect of 'EDUCATION'. I spent the whole of that evening assessing my chances of obtaining this priceless jewel. My study in the subject of mining had included several branches of learning which I knew would be useful and already I had entered the fascinating world of 'let x equal'. Our Parish Priest had taken note of my interests and had helped me in my effort to understand Latin Grammar, and John Spode, being a lay–preacher, had made sure that I always had a good book in my hands. All this, together with a quiet, warm room at Silverdale Vicarage gave me the taste of the luxury for which my heart and mind craved. Soon I turned my back on the depression and atmosphere of utter hopelessness and travelled to the south coast to take up this offer which the advertisement had announced.

Years of study lay before me, but it all had to be done after my day's work—evening classes, reference rooms in public libraries. The majestic works of Shakespeare, Palgrave's Golden Treasury and a thousand other books were my constant companions over the years. I turned over the pages and never lost sight of my good fortune at having access to such riches. The Workers Educational Association urged me on and provided means for improvement and even raised my hopes of a scholarship to Ruskin College, Oxford, but simultaneously with this prospect came an opportunity to enter a Teachers' Training College. Full–time study at last; I grasped it with eager hands.

Standing before my first class of children as their teacher I suddenly became aware of my responsibility and soon a delicate rapport developed as I realised that it was the children who were teaching me. Then came my first permanent appointment among the sturdy boys of Kent and within months I had organised a party of them who were of the permitted age, to visit the Chislet Colliery and spend the day underground. At once I knew that those five years as a miner were now a vital experience in my life and every growing child who passed through my hands received some instruction in the subject of coal–mining and the character of mining people. The words of my father and Fred Askey came ringing back into my consciousness, 'Remember Harold . . . LIVES ARE IN YOUR HANDS.' Once I was fully qualified and my certificate endorsed, doors were flung wide open for me to attend further courses in every branch of education.

After six years of married life we were blessed with a son. Immediately he was old enough I took him to Staffordshire to learn about his father's kinsfolk. I watched his growing interest in the pit and the miners who came up the shaft with black faces, shouting across to me, 'Ah–do, 'arold, costna keep away from th' pit then?'

Once in his teens my son bullied me to take him down the pit and through the generosity of the National Coal Board and the pit General Manager I again stepped into the cage and descended my old pit, but this time with my arm around the shoulder of a young man, my own son. That new pit–bottom with its high walls had closed in. The weight had pushed downwards and inwards and the cathedral–like appearance had disappeared; nature had made its mark on the efforts of man. I pointed up to the spot where Keele Dip engine had been housed and said to my son, 'there it is David lad; that is the place where my engine hauled the tons of coal up Keele Dip.' After a most thrilling day during which we visited the coalface we stepped off the cage and visited the modern bathhouse and washed off our pit–dirt. Day–shift men were coming off, afternoon–shift men changing clothes to go down.

A loud voice rang through the bathhouse as we dried ourselves. 'Eh–up, Well! Well! Ah–do Harold, didstna thaiy see enough of this pit?'

It was the cheerful voice of Dick Ward. He was in my class at day school and he was always lovingly known as Dickie Ward, for he was a happy, good natured boy.

With a slight tone of pity in my voice I replied to his greeting, 'Hello Dick lad. You are still swinging at the end of this pit–rope then?'

'Ah, Harold, I am, but I've only another nine years to go now. They wunna see me up anywhere near this pit once I've retired I can tell thee. I've seen thee up around this pit–bank many a time since thou hadst th' brains to get thyself out of it. Dost remember Harold, doing two jobs for a fortnight so that thou couldst work thy notice and please thy faither? Costna keep away Harold?'

'It is something you would not understand, Dick, you never having been away from it.'

We had been invited to use the canteen facilities and Dick Ward seemed unwilling to leave us. He was changed and ready to go down the pit, but he accompanied us and sat with us as we partook of refreshment, provided by the Management. Smilingly he nudged me and said, 'Thou wast clever, Harold, to get out of th' pit.'

'Clever you say Dick. Let me explain.' I reminded Dick Ward of all the circumstances and the conditions of those far–off days the thousands on the dole, the hunger, depression and stress. My son's whole being seemed to dance with excitement as he looked around and watched the miners and listened to their chatter.

'Look Dickie, I suffered drastic reduction in wages when I left this pit, but my highest price was the fact that I found myself no longer working among miners. There is a "something" down that pit among those men which both you and I have experienced and it defies description. It grew on me during those five years and you have had it for a whole lifetime. That, Dick, is my answer to your question as to why I am to be seen on this pit–bank every time I visit Silverdale. It seems just as though I am trying to get a bit of that quality, that relationship rubbed back on to me.'

'Well Harold, once clear of the pit and into retirement I shall spend my time down Back Lane, in the Bluebell Wood, over the Bogs Fields and away to Keele; I shall spend some of my time up on Spring Bank taking in a sample of that clean air into my lungs, that air that thou wast always on about. Dustna think, Harold, that I've taken enough of that pit–dust into my lungs since I was a kid of fourteen. Remember!? For years now I've counted the days when I should not have to hang on the end of a pit–rope every day and be lowered into darkness. Dostna think that I've earned a bit of fresh air?'

A pang of compassion brought a lump into my throat as he turned to go. Dick Ward noticed that my son was looking up to his face and taking notice of his pit–clothes and he said to my son, 'Never, never come to wear these lad.'

As cheerfully as my emotion would permit I said, 'I hope Dick that these next nine years will fly by quickly for you.'*

My son David looked for an opportunity to speak and once I had turned to him he asked, 'How long does a miner have to tolerate that life?'

'You heard what Dick Ward said, "Only another nine years to go now". In nine years he will be sixty–five. He started in the pit with me at fourteen. Fourteen from sixty–five equals fifty–one years underground. That's a lifetime. That's too long for any man to spend in darkness.'

'Goodness, all those years shut away in that dark place, taking in foul air; why, people who have never seen a pit know nothing of the miner's life do they? I do not suggest that I know anything after one quick look round, but I *did feel* the danger around me, just as you have often described it.'

'What you saw down that pit does not compare with the many aspects of danger which existed in my day. Danger simply shouted at the miner from every corner. There was much less space than there is

*Dick Ward was killed underground not many months after the day of this conversation.

177

today, but even now everywhere is cramped. Remember what that face–worker said to you, "We get to know each other very well down here young Mr Brown." One day you will tell me what impressed you most, what were your deep reactions.'

We had one whole clear day left before our return to the south coast. We put this day to good use by sauntering through the streets of Silverdale, then into Back Lane. David stopped at every place of worship and gazed, but without any word. The same happened in Abbey Street before the house where I had been brought up. I did not disturb his reverie. We came to Burn Spout, a gushing spring of water from the hillside which had gone on for many years. Standing there trying to understand its never–ending flow David said, 'Father, let's walk around Keele.'

Slowly we climbed the gradient of Rosemary Field and threw ourselves down under the very tree where I had gazed at the twinkling lights of the Potteries towns and battled within my mind to find some means of escape after only one day down the pit.

'Complete darkness, father, except for the tiny flame of an oil–burning lamp.'

'Yes lad, that's what it was years ago and that point alone once made me try and assess the price a miner pays for coal to be produced. In this age adequate compensation would be paid if someone knocked you off your bike or injured you in any other way, that is if neglect could be proved. Now, supposing you were deprived of fresh air, light, space around you and, above all, the warmth of sunshine and all its healing benefits and that you suffered this loss for the space of a third of every working day of your life; all this loss due to the negligence of another person, what sum of money would compensate for such a loss?'

'Goodness! How could such a sum possibly be computed? It hardly bears thinking about.'

A period of silence followed. Then I reminded David, 'Now you see, the miner not only loses all those pleasant amenities of life when underground, he also has to tolerate all the features which seemed to have shocked you after your visit to the face, groping about in semi–darkness, breathing in foul air, hot conditions, no toilet facilities, terrific weight overhead and awareness that the pit could blow up at any moment.'

We reached the village of Keele, turned right into the unmade

Quarry Bank Road and soon, away on our left, we could see the wide Cheshire Plain stretched before us. My eyes travelled no further then the prominent hill of my boyhood and years of youth. Not a word was spoken for some minutes; then, 'Father, as we sat silently under that tree, top of Rosemary Field, I thought to myself what a wonderful thing it would be if the attitude and conduct of miners, the quality of behaviour which you have often described to me, could be brought up on to the surface and implanted into society everywhere, retained there for all time and accepted as the rule.'

I looked at his face with amazement and with some degree of pride, 'Aye lad, it would, it would be wonderful, an aspect of Utopia I should think.'

I think my son David saw that my gaze remained on the hill, Spring Bank, and he respected my privacy during the silence of that moment. At length I asked, 'David, anything else on your mind?'

'Yes, you asked what it was impressed me most at the pit. I would like to know what impressed *you* after so many years away from pit–work.'

It was not an easy question to answer for the whole visit had been a wonder to me. 'Quite a lot impressed me, enough to fill a book, but what stood out was the obvious SAFETY EVERYWHERE. The Safety Officer to start with, his dedication. As I listened to their remarks I noted how "THE SAFETY OF THE MEN" and their welfare were of paramount importance. Thank God for their effort and their dedication and for all those who insisted upon it all coming to pass. All that, David, is a wonderful thing to me. Pits are safer now and what a pleasure it would give to your grandfather if he could see it all. If only he could see the improved roads, the hot bath so that men can now leave their dirt at the pit, then, at weekends, take home a wage commensurate with their effort and skill. The miner's skill has never been fully recognised by the general public. They are craftsmen and because of their inherent courage and endurance, and the noble qualities they display underground, I have always been proud to be associated with them.' We both again fell into silence, lost in our thoughts.

'Father,' said my son slowly, 'It is amazing how my ideas about a pit and miners have changed since I stepped into that cage and dropped into the darkness.'

I was pleased with that remark for it epitomised so many of my own ideas. 'Yes, of course. You have seen it all for yourself, you will now think so much differently, change your standpoint, see things from a different angle. Just pause for a moment in your thought for there is another relevant matter. Your ideas about miners have changed in twenty four hours or so, due to the fact that your experience has been extended. Throughout the long years that I have studied, read far into the night, attended courses and listened to endless lectures on radio, I have found that my ideas were constantly changing, almost with every book I read, certainly with every year that passed. But there is one thing which has never changed, it has never varied in the slightest degree, it has remained firm as a rock.'

My son's excitement became manifest in his features. 'You see, my son, it is easy to see how my opinions could have changed and my ideas varied when I moved into fresh areas, mixed with different classes of people and engaged in various types of occupation. Since 1927 when I left this pleasant place I have moved among so many different social levels, I have compared so many different attitudes to life and culture. I have travelled widely and all the time I observed, watched, took notes. Naturally, my ideas have changed.'

David showed some impatience, but slowly I drew him to my relevant point. 'The one thing that has never, never varied has been my appreciation of the character of the miner. I have firmly established in my mind that my experience underground among these men helped me to consolidate my "TERMS OF REFERENCE". When at any time I have been obliged to assess character or performance, the standard set by those miners has always been my measuring–rod.'

'Yes, father, I can't forget the words of that miner to me at the face and each time you have brought me up here among them I have felt the warm, friendly atmosphere and, just as you have described it to me, it does all seem to seep through into their lives on the surface.'

'David, what makes a soldier display bravery in the trenches, the Lifeboatmen out on dangerous operations, men in submarines? The same urge is built into the miner's character all the time he is underground; constant danger creates responsibility within a man and it is difficult to translate it into simple language.' David paused, but he came nearer and I could see that he wanted to say something.

'Well?' I asked, 'what is it?'

'You know father, you have never really left that pit have you?'

'I thought I'd escaped in 1925, but as the years mounted up I realised that what I had seen underground far exceeded anything I had encountered on the surface. I am firmly of the opinion that the history of pit–disasters in our land and the response of rescue teams and other volunteers is an emblazoned record of herosim which has never been fully recognised.

'I heard you say that some years ago, father.'

'Yes, and I now have to find fitting language to describe it; the intensity of self–sacrifice, the display of the highest code of conduct when faced with danger and tried to the limit of endurance.'

'What a pity you did not make notes of your experiences. Writing the book you often talk about would now be an interesting task for you.'

'Not much has been lost for I have relived those pit–years over and over again and I have retained the memory of the way a man's character was bound up with the way he did his work, how the measurement of his worth in society was assessed in accordance with his attitude toward his mates underground when danger threatened, and this applies from the General Manager down to the lowliest waggoner. A man is not dishonoured down the pit because he peforms a humble task; no job down there is frowned upon. A man is a miner and that fact alone raises him to a place of respect in the eyes of mining people. Every man who goes underground accepts a certain degree of responsibility for the safety of the whole pit. What remains so vividly in my mind throughout all these years is the complete absence of deceit; I remember well that there was no lip–service nor any corresponding pretence or sham of any kind. There was no place down the pit for these items of flimsy veneer; as I have already said, there was no audience, no–one to applaud, no–one to impress except those men who knew good workmanship from bad. Only a man's worth was recongnised, how he maintained the traditional standards underground. There was room only for "real men". My observations told me that it mattered not to what layer of society of man belonged, it mattered not in what country he had been born nor even what the colour of his skin was, for, within the hour of the commencement of the shift, the bodies of all colliers were black. David, lad, what captured my heart and mind was the way they behaved when under stress; the memory of that alone is the impulse which brings me up here on the slightest excuse. Those miners taught me to cultivate a masterful attitude of mind in order to be able to meet adversity without losing equilibrium; a dozen names come to mind as I utter those words, there are thousands like them in other mining areas. Their example taught me that being lowered into the earth to earn one's living did not mean that a man had to lose his dignity.'

There was a long pause at the end of which David asked almost nonchalantly, 'Father, what was the attitude of the womenfolk to all that hardship?'

'Oh dear David, *there's* a question which cannot be fully answered in a few sentences, but I'll try. Their excellent qualities can only be appreciated by actually seeing the anxiety and concern written across the countenance of a miner's wife or mother. They shared every hardship whether with husband, member of family or community. Their apprehension included the risk and danger of the work underground and their concern extended to the stringent economy which had to be exercised, year in, year out, in order to make ends meet. You'd know what their attitude was if you could witness just once the expression of humiliation on the faces of those in the procession to the pawnshop on Monday mornings. See that, and you begin to appreciate the extent of their struggle for existence. Those women occupy the highest place in my esteem. I owe them a debt for often they took my hand and pointed the right way for me to take. To feel oneself a member of their society, David lad, is a privilege. Already you have identified yourself with them and I trust that if you have a family of your own, you will teach them about these North Staffordshire people from whom you have sprung. It will be no easy matter for you to maintain their high standards and traditions, but if you do, you will always be able to hold your head high and feel proud. Underground and on the surface men and women sacrificed personal comfort and leisure in the interests of common good and that in my opinion, is a sound basis for any well–ordered society. My memories of this phase in the history of our local community are vivid. I was there when these conditions were in existence. Short–time working at the pits and in other heavy industry, the general low standard of wages and the corresponding light purses, all brought hard times and continued stress which called for the personal sacrifice I have just mentioned. Nevertheless, because of these acute circumstances, because of the ready response of people to help each other, life became enriched. The grandeur of their sense of responsibility towards each other brought a lustre into our community life. We were a happy society making the best of difficult times. I have never forgotten the feeling which it all gave to me in my boyhood days.'

I went on to describe how miners used to say that working in cramped positions down the pit contributed to the miner's urge to burst forth into song once he was back into fresh air. When these cramped positions and the stress of poverty were too much to bear, it created a kind of mental cramp. Bursting spontaneously into song seemed to bring some inner relief, hence the tradition among miners for high performance in choral works. One miner once told me that he suspected that it was the same psychological force at work in the ponies when they were brought up from the darkness and turned into the park–field for one short week at the annual Wakes Week; they displayed such joy as they ran about.

I stopped talking for I had poured out my feelings. My son caught me gazing across at Spring Bank. He came and stood right beside me and placed his hand on my shoulder. He smiled as he remarked, 'There it is, your shrine, father, your place of memories. I remember the first time you took me to the summit.'

I returned his glance and wished that my own father could have shared that moment. As I admired the attractive clump of trees on the hill, standing like a sentinel guarding the vast plain below, I turned to my son again and with words which matched my mood, 'I would like to have the opportunity to turn to all those men and women who shared those days of struggle and say to them, "Thank you for your excellent example and your good nature". But could I find suitable words to convince them of the depth of my gratitude?'

'If you wrote your book, father, that would give you the opportunity to express your tribute to them and relate your sincerity.'

Arriving back home on the south coast, I went straight to my notes which I had made over the years for I knew that the time had come for me to begin feeding the endless sheets of paper into my typewriter.

Index